Exploring Science

4

MANOJ PUBLICATIONS

Exploring Science-4

Publisher:
MANOJ PUBLICATIONS
761, Main Road, Burari, Delhi-110084 (INDIA)
Mobile : 09999476076, 09868112194,
08178823569, 08178854810
Email : info@manojpublications.com
for online shopping visit our websites:
www.sawanonlinebookstore.com

ISBN : 978-93-5579-334-8

Concept by:
Sanyam Gupta

Edited by:
Rohan Kumar

Preface...

The series entitled Exploring Science comprises a set of five books from classes I to V. Based on the NCERT syllabus, the series follows an activity-oriented thematic approach in which the child's learning is assessed in a comprehensive range of situations and environment both in and out of the classroom.

Each book of the series is divided into different topics. Each chapter starts with a synopsis of topics discussed under the heading Turning Points. Rack Your Brain is an explanation that makes the child feel aware about the topic and relates it to his daily life situations. The content of each chapter is developed in a simple and lucid way, and enriched with colourful illustrations. At the end of each chapter Summary of the Chapter is given to recapitulate the chapter.

The Experimental Work under the heading Get Hands-on Experience and Hands-on Learning is carried out in the form of Classroom Assignment, Home Assignment, Project Work and Classroom Presentation. This will help to link the textual knowledge with the practical experience.

We have made our sincere efforts to make this series error-free, refined and concise. The worthy suggestions for its further improvement are most welcome.

Contents...

1 Healthy Food and Health

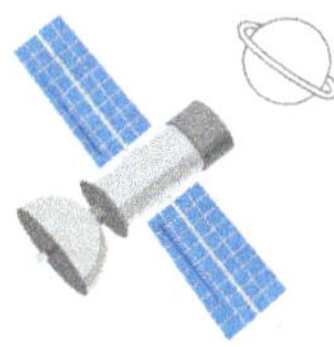

We Will Learn

- Kinds of Foods
- Balanced Diet
- Keep Your Body Fit and Stay Healthy

Rack Your Brain

Food is one of our basic needs. In our day-to-day life, we do many activities like playing, studying, dancing, swimming, etc. To do all these activities , we need a lot of energy. Food gives us the energy to do activities. It also protects our body from many diseases. In this chapter, we will learn about the main nutrients of food and their uses.

Kinds of Foods

A food group is a collection of foods that has various nutrients. Nutrients are the substances that are needed by our body for good growth and health.

Types of Food Nutrients

- Energy-giving food → Carbohydrates and Fats
- Body-building food → Proteins
- Protective food → Vitamins and Minerals

Along with these, our food also contains water and roughage which do not have any nutritional value but help in the proper functioning of our body.

We should eat the food that has all these essential nutrients. These nutrients give energy to our body. They help our body to repair the damaged cells. They also protect us from falling ill very often.

Carbohydrates

Carbohydrates are the main source of energy. They are needed by our body. They are called energy-giving foods. When consumed in excess, they are converted into fat and stored in the body. The deficiency of carbohydrates in the diet does not occur except in times of food shortage. Food-items like fruits, vegetables, cereals and bread are the sources of carbohydrates.

Fruits Vegetables Cereals Bread

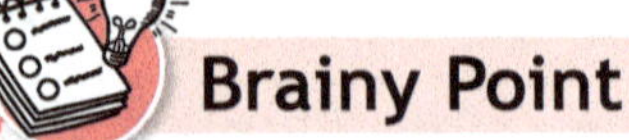

Brainy Point

Starch and sugar are important carbohydrates.

Proteins

Protein is an essential nutrient that is needed by our body for the growth, repair and building of body tissues. Hence, proteins are known as body-building foods. The deficiency of proteins in the diet of adults results in the loss of weight, reduced resistance to infection and oedema. Food-items like milk, cheese, fish and pulses are the good sources of proteins.

Eggs Milk Pulses

Fish Meat Cheese

Brainy Point

Lack of protein in the diet causes a disease called 'Kwashiorkor'.

Fats

Fat is an essential part of our diet. Fats are also energy-giving foods. We need fat for our nerves, brain and skin cells to protect vital organs in the body and to help to control our body temperature. Our body needs a little amount of fat. Too much fat is unhealthy for us. Food-items like oil, butter and nuts are the good sources of fats.

Butter Oil Ghee Nuts

Vitamins and Minerals

These are present in very minute amounts in different foodstuffs and are essential for growth and health. Vitamins and minerals also protect our body from diseases and keep us fit. They are known as protective foods. The more important minerals are calcium, phosphorous and iron. Food-items like fruits, vegetables, milk, eggs and peanuts are the good sources of vitamins and minerals.

Peanuts Milk

Fruits

Vegetables

Egg

Both calcium and phosphorus form the major constituents of bones and teeth and are essential for their formation. A deficiency of calcium in the diet causes rickets and decay of teeth. Food-items like milk, curd, almond and cheese contain calcium. Iron is mainly required for the formation of haemoglobin which is an important constituent of the red blood cells. Prolonged deficiency of iron in the diet causes anaemia. Food- items like green leafy vegetables, beetroot, apples and carrots contain iron.

Roughage and Water

Roughage, also known as fibre or bulk, is an indigestible compound that your body can't absorb. It is found in many fruits, vegetables, salads, grains and legumes. It is very important for our body as it helps us to remove our body waste.

More than half of our body-weight is water. It is an important part of our food. We cannot live without it. It helps in the digestion and absorption of food. It also helps to regulate the temperature and the activities of our body. We should drink plenty of water every day.

Fruits

Vegetables

Cereals

Water

Brainy Point

Nuts and seeds are quick snacks and a wonderful source of fibre.

Balanced Diet

A balanced diet is the one that contains all the nutrients in proper amounts. It helps maintain or improve overall health. It gives our body all the nutrients it needs to function properly. In order to get truly balanced nutrition, we should obtain the majority of our daily calories from fresh fruits and vegetables, whole grains and beans.

Keep Your Body Fit and Stay Healthy

As we eat healthy food to keep our body fit, similarly we should also give rest to our body. For example, we should sleep for at least 6-8 hours every day. If we do not take enough sleep then our body will not function properly. Hence, we will fall ill.

Correct posture is also essential for our healthy body. While standing, our body should be erect and straight. When we sit, we should keep our back straight. If we sit or stand in the wrong position then it will lead to pain in our legs, arms and joints.

Running, walking and gardening are good for our body. Regardless of what you do, regular exercise and physical activity are the path to health and well-being. Exercise burns fats, builds muscles, lowers cholesterol, eases stress and anxiety, and lets us sleep restfully. We should also play games like cricket, basketball, etc.

Sleeping child

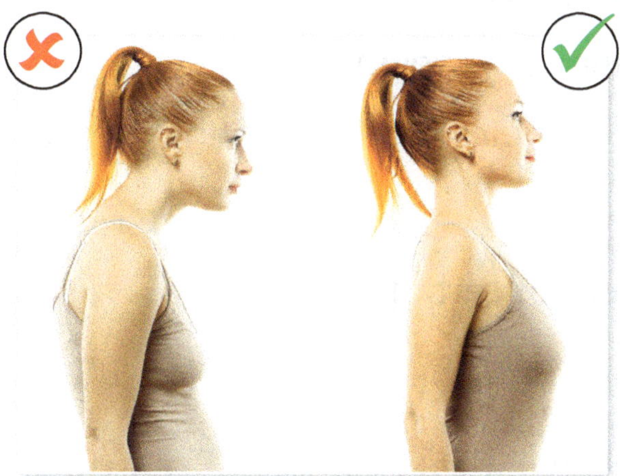

Correct posture

Exercise

Summary of the Chapter

- Food is the fuel for the body. It helps us to grow, gives us energy and keeps us healthy and warm.
- Carbohydrates and fats are called energy-giving foods.
- Vitamins and minerals protect our body from diseases.
- The diet which has all types of nutrients in the right amounts is called a balanced diet.
- Water helps to regulate our body temperature.

A. Answer the following questions :

1. What is food? Why do we need food?
2. What are nutrients ?
3. Name the various nutrients needed by our body.
4. Why are carbohydrates necessary for our body?

5. What are protective foods? Give examples.
6. What is roughage?

B. Fill up the blanks :

1. Iron is mainly required for the formation of ______________ .
2. Vitamins and minerals protect our body from ______________ .
3. Fish and meat are preserved by ______________ .
4. More than half of our body-weight is ______________ .
5. ______________ are required for promoting growth.

C. Give one-word answer :

1. Write any two sources of carbohydrates.

____

2. Name a source of protein.

____

3. Which nutrient is not digested by our body?

____

4. Name a source of iron.

____

5. Which nutrients are known as body-building foods?

____

D. Write (T) for a true statement and (F) for a false one :

1. Food gives us nutrition/materials. [ ]
2. Potatoes are the source of proteins/carbohydrates. [ ]
3. Pulses are the source of proteins/roughage. [ ]
4. Fat/Water keeps our body warm. [ ]
5. Vitamins/Fats are known as protective foods. [ ]

Make a list of food-items that you should avoid eating or eat in very small amounts:

Names of foods that should be avoided

_______________ _______________ _______________

_______________ _______________

Names of foods that should be eaten in small amounts

_______________ _______________ _______________

_______________ _______________

Make a chart showing a balanced diet for a school-going student. Display it on the school bulletin board, and spread awareness about the need for keeping healthy and staying fit.

2 Care of Teeth and Microbes

We Will Learn

- Temporary and Permanent Teeth
- Types of Teeth
- Structure of Our Teeth
- Taking Care of Teeth
- How to Brush Our Teeth
- Microbes

Rack Your Brain

Today, Golu was very happy. He was very fond of sugarcane; his father had brought a sugarcane for him. He went to his Grandpa and said, "Grandpa, after having lunch we will eat the sugarcane." Grandpa said, "I cannot eat the sugarcane because my teeth have fallen off." Golu said, "Don't worry; I have seen a film about artificial teeth on Discovery Channel. I will tell you all about them."

Dear children! Teeth play an important role in our daily life. They help us to bite and chew the food we eat. Teeth can be healthy if they are taken good care of. In this chapter, we will learn about the types of teeth, care of teeth and different types of microbes.

Temporary and Permanent Teeth

Our teeth are found in our mouth. Teeth are composed of calcium, phosphorus and other minerals. When we were babies, we didn't have teeth. Teeth started growing when we were about six months old.

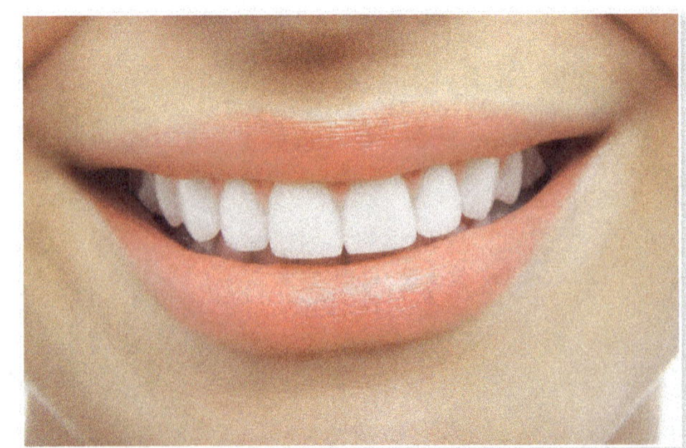

Most kids have their first set of teeth by the time they are 3 years old. These are called the milk or baby teeth. There are 20 milk teeth in a baby's mouth. When a child is about 5 or 6 years old, milk teeth start falling out, one by one. Slowly, the permanent teeth grow in and take the place of the milk teeth. By the age of 12 or 13, most children have lost all of their baby teeth and have a set of permanent teeth.

There are 28 permanent teeth in all. Between the age-group of 17 and 25, four more teeth called wisdom teeth usually grow at the back of the mouth. The wisdom teeth complete the adult set of 32 teeth.

Types of Teeth

There are different types of permanent teeth in our mouth. Each one has its own function. Let us read about them.

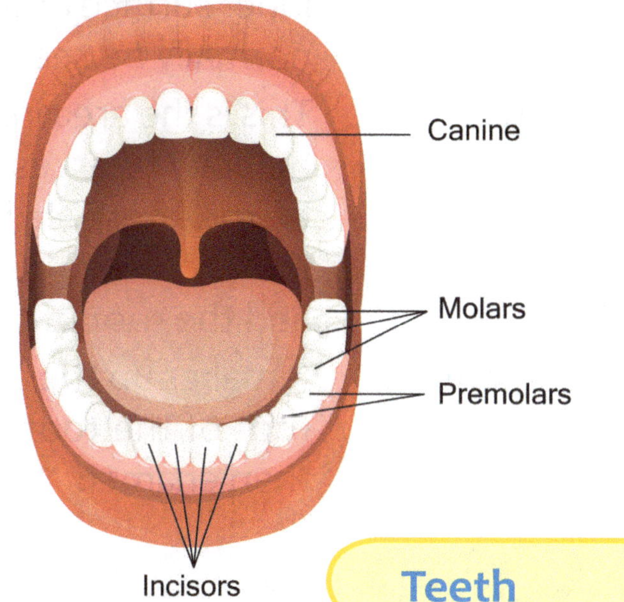

Teeth

Incisors : Incisors are eight in number. There are four incisors in each jaw. They are used for cutting and biting food.

Canines : Canines are four in number. There are two canines in each jaw. There is one canine beside each incisor. They are used for tearing food. Canines are sharp teeth.

Premolars : There are eight premolars, four in each jaw. They are beside the canines. They are used for crushing and chewing food.

Molars : Molars are twelve in number. These are six in each jaw. They are big and flat. There are two molars on each side of the premolars. Out of twelve, only eight molars grow first and the remaining four grow later. They help to chew and grind food well.

Brainy Point

It was only about a 100 years ago that someone finally created a cream to clean teeth.

Structure of Our Teeth

Each tooth in the mouth contains four different tissues that serve different functions. The teeth are made up of two major parts:

- The crown
- The root

The crown of the tooth is what is visible in the mouth.

The root of the tooth is the portion which normally is not visible in the mouth and is inside the gum.

Within each tooth, the four different tissues that are present are enamel, dentine, pulp and cementum.

Enamel : It makes up the protective outer surface of the crown of the tooth.

Dentine : It makes up the majority of the inner surface of the tooth. It cannot normally be seen except on x-rays.

Pulp : This is the area inside the tooth that holds the nerves and blood vessels of the tooth. It is in the centre of the tooth and is in both the crown and the root of the tooth.

Cementum : It makes up the outer surface of the root of the tooth. It is much softer than enamel.

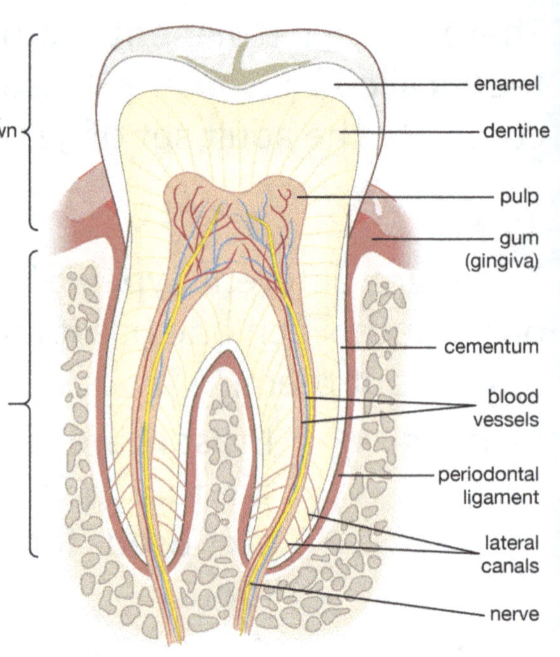

Structure of Tooth

Brainy Point

Tartar (yellow film) collects on the teeth. This makes the teeth look unclean and makes the gums weak.

Taking Care of Teeth

Taking care of teeth is necessary to be healthy. If our teeth have any problem then we will have problem in eating food. Let us read about some points through which we can take care of our teeth.

- We should brush our teeth at least twice a day. It is very important to brush teeth before bedtime.
- We should rinse our mouth after every meal.
- We should eat healthy food that makes our teeth and gums strong.
- We should eat hard fruits like apples as they make our gums and teeth strong.
- We should not eat too much sweets as they stick in between our teeth and spoil them.
- We should visit the dentist once every six months for the check-up of our teeth.

How to Brush Our Teeth

The best way to brush your teeth is in little circles. Go around and around until you have covered the surface of each tooth. Brush up and down, rather than side to side.

You can also clean between your teeth with dental floss at least once a day. Dental floss is a special string used for cleaning your teeth. It helps to remove food and plaque that get stuck in between the teeth. You can also brush your tongue to help keep your breath fresh!

Always remember: if you take care of your teeth now, you'll be chewing like a champ for the rest of your life.

Microbes

The germs that enter our body are so small that they can be observed only under a microscope and hence they are also known as microbes.

There are many types of microbes.

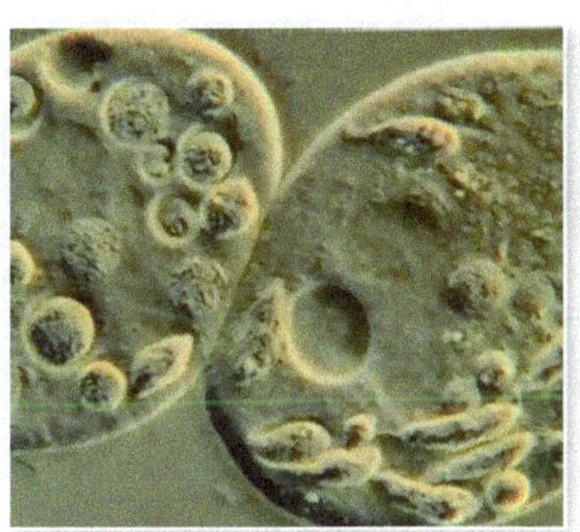
Protozoa

Fungi

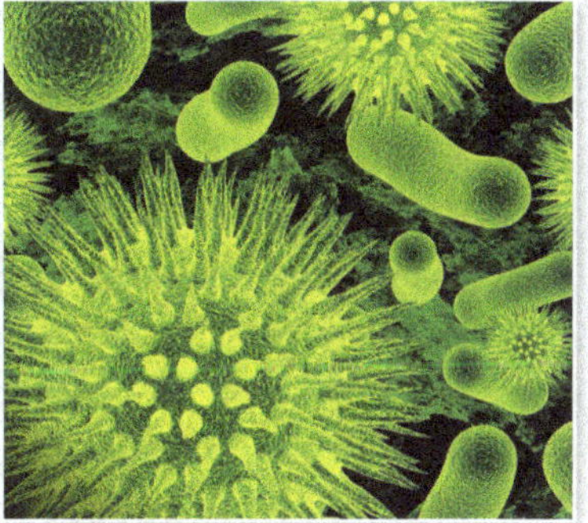
Viruses

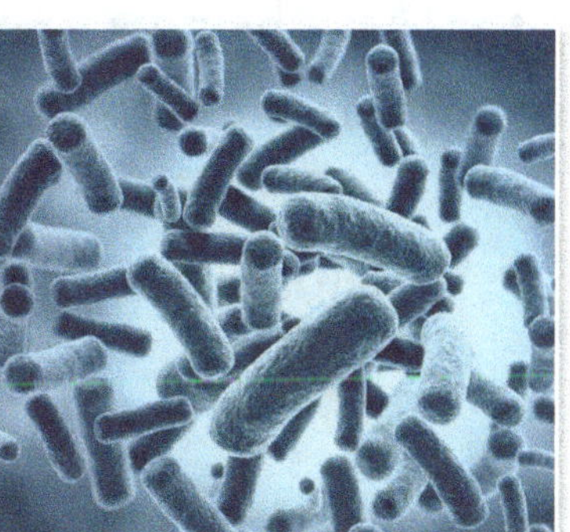
Bacteria

Bacteria : These are micro-organisms that are actually made up of single cells. They cause diseases like typhoid, tuberculosis, cholera and pneumonia. Some of these bacteria help us make curd from milk and some medicines.

Viruses : These are the smallest of all microbes which cause diseases like chickenpox, smallpox, influenza, common cold and polio.

Fungi : These are non-green plants that grow on dead matter. They cause skin diseases like ringworm. They are also responsible for food poisoning resulting from eating stale food.

Protozoa : These are single-celled animals that grow in dirty wet places. They cause diseases like malaria, dysentery, etc.

Useful Microbes : All microbes are not harmful to us. In fact, some microbes are useful to us as they help to produce things that we use.

- Some bacteria help to prepare curd. Wine and vinegar are also made with the help of bacteria. Bacteria also help to break down food into glucose.
- Dead plants and animals are broken down to a simple form by some bacteria and help to keep the environment clean.

- Medicines like penicillin are made from bacteria.
- Some fungi like yeast are used for preparing bread, idlis, cakes, etc.
- Mushrooms which are eaten as a delicacy are also a type of fungus.

Brainy Point

Our mouth is a home for more than 500 types of microbes. Hence, keeping our mouth clean is very important.

Summary of the Chapter

- Teeth help us to chew food and also help us to talk.
- There are two sets of teeth-milk teeth and permanent teeth.
- We should take care of our teeth by brushing them after each meal.
- Microbes are tiny living things that are found everywhere.

A. Answer the following questions :

1. What are milk teeth and permanent teeth?
2. What are the different kinds of teeth? For what purpose are they used?
3. What are the parts of teeth? Define.
4. How can we keep our teeth clean?
5. What is a dental floss? What is its use?
6. What are microbes?
7. What are the types of microbes?
8. Write any three uses of microbes.

B. Fill up the blanks :

1. Iron is mainly required for the formation of ____________ .
2. There are ____________ milk teeth in a baby's mouth.
3. The ____________ teeth complete the adult set of 32 teeth.

4. ______________ are used for tearing flesh.

5. ______________ makes up the protective outer surface of the crown of the tooth.

6. All microbes are not ______________ to us.

C. Write (T) for a true statement and (F) for a false one :

1. When a child is about 5 or 6 years old, milk teeth start falling out, one by one. ☐
2. The milk teeth grow in and take the place of the permanent teeth. ☐
3. An adult has 33 teeth in all. ☐
4. Premolars are used for crushing and chewing food. ☐
5. The root of the tooth is the portion which is visible in the mouth. ☐

D. Give one-word answer :

1. How many milk teeth does a baby have?

____

2. How many wisdom teeth does an adult have?

____

3. Which part of a tooth holds the nerves and blood vessels?

____

4. What is the inner surface of a tooth called?

____

5. Name the hardest material in our body.

____

6. How many cells make up a bacterium?

____

Get Hands-on Experience...

Visit a dentist's clinic with your parents and collect information on the care of teeth. Write the ten lines of your experience :

Hands-on Learning...

Draw and label the parts of the tooth:

3 Digestion of Food

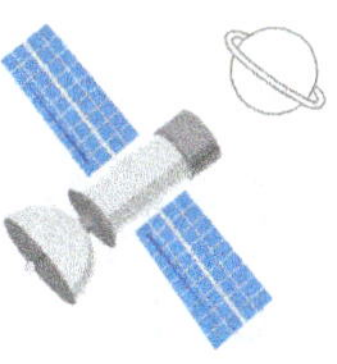

We Will Learn

- Why Do We Need to Cook Food ?
- The Best Way to Cook Food
- Preservation of Food
- The Journey of food
- Some Rules for Healthy Eating

Rack Your Brain

Aman and Tarun are good friends. They go to school together. Today, Aman is absent because he has got fever. The teacher asked Tarun about his friend. Tarun said, "Ma'am, Aman always likes to have fast food and aerated drinks. He hates to eat green vegetables and fruits. He does not eat a balanced diet. So, he felt sick and could not come to school."

Dear children! Fast foods and cold drinks contain only carbohydrates. They have no nutrients in them. If you want to stay healthy, you should eat balanced food-items that contain different nutrients. Green vegetables and fruits must be included in our daily diet. These eatables keep us away from diseases. Let us read about the food needed by our body.

Why Do We Need to Cook Food ?

Food is cooked to–

- make it soft and easy to chew and digest.
- make it edible and tasty.
- kill harmful germs.

The Best Way to Cook Food

- Vegetables should be washed before cutting.
- Cook the food in just enough water to retain nutrients. The excess water should not be thrown away as this results in the loss of nutrients from the food.
- Overcooking the food destroys nutrients.

Preservation of Food

If we keep food-items for a long time, they get spoilt. Spoilt food is unhealthy for us. We should not eat spoilt food. It can make us fall ill. So, in order to keep food-items fresh, we can preserve them by various methods. Let us read about some methods.

Refrigerating

We should keep food in the fridge to keep it fresh for a few days. For example, we keep the leftover cooked food in the fridge and use it the next day.

Drying

Some food-items can be stored after drying them. For example, grapes are dried for making raisins.

Pickling

Fruits and vegetables are mixed with oil and salt to make pickles. Adding salt to pickles makes them stay fresh for a few days.

Canning and Bottling

We make sauces and jams and preserve them in cans and bottles.

Deep Freezing

Some food-items are deep-frozen to preserve them for a long time. These food-items can be stored for months. For example, fish, meat, corn, peas, etc.

The Journey of Food

Our body needs food to provide it with energy, vitamins and minerals. However, in order to use food, we must first break it down into substances so that various organs and cells in our body may use it. This is done with the help of our digestive system.

The digestive system acts in stages to digest our food. Each stage is important and prepares the food for the next stage.

Here are the major stages of the digestive system:

Chewing : Chewing is the first stage of digestion. When we chew our food, it breaks down into smaller pieces. Smaller pieces are easier to digest and swallow. There is saliva in our mouth which mixes with the food and helps in digestion.

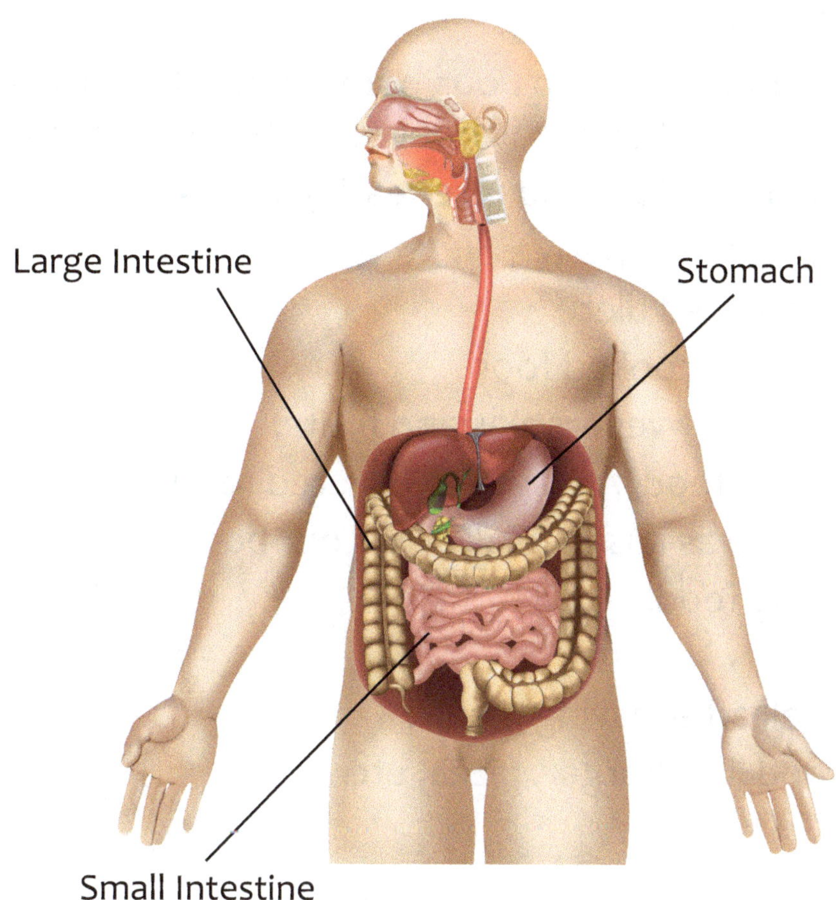

Digestive System

Swallowing : After the food is chewed, it is swallowed in the stomach. Our tongue helps to push food into the back of our throat. Then, there are special throat muscles that force the food down into a long tube that leads to our stomach. This tube is called oesophagus or food pipe. The muscles of the food-pipe push food down into the stomach.

Stomach : The next stage happens in the stomach. Food remains in the stomach for about four hours. While the food stays there, more enzymes go to work on it, breaking down things like proteins that our body can use.

Small Intestine : The first part of the small intestine works with juices from the liver and pancreas to continue to break down our food. The second part is where the food gets absorbed from the intestine and moves into the blood.

Large Intestine : The last stage happens in the large intestine. Any food that the body doesn't need or can't use is sent to the large intestine and later leaves the body as waste.

Some Rules for Healthy Eating

We should eat healthy food as it keeps our body fit and fine. Healthy food habits keep us away from diseases. Let us read about some healthy eating habits.

- Always wash your hands before and after eating meals.
- Chew the food slowly and properly.
- Always try to eat a balanced diet.
- Always eat fresh and well-cooked food.
- Try to eat more green leafy vegetables and fruits.
- Avoid eating uncovered food. Always eat covered food.
- Avoid eating street food as it is unhealthy for our body.
- Never talk while eating food.
- Never overeat.
- Rinse your mouth well after having meals.
- Avoid eating junk food like, burger, pizza, etc.

Summary of the Chapter

- The body needs a diet which has different types of nutrients which are present in the different types of food-items.
- The variety of food we eat is broken down into simple form to be digested. This is called digestion.
- A balanced diet supplies the right amounts of carbohydrates, fats, proteins, vitamins and minerals.
- Cooking makes the food tasty, soft and easy to digest.

A. Answer the following questions :

1. Which method is used for preserving the leftover cooked food ?
2. Name the various methods by which we can preserve food.
3. Why do we need to cook the food ?
4. How does the food reach the stomach?
5. Write any two points of healthy eating.

B. Fill up the blanks with the suitable words :

1. We eat the food in both the ways, as ____________ and ____________.
2. ____________ and ____________ can be eaten raw.
3. Grapes are dried for making ____________.
4. Our ____________ helps to push food into the back of our throat.
5. Chew the food ____________ and ____________.

C. Tick (✓) the correct answer :

1. Food is cooked to make it hard ☐ / soft ☐.
2. Overcooking the food destroys ☐ / saves ☐ nutrients.
3. Adding salt ☐ / sugar ☐ to pickles makes them stay fresh.
4. The food is chewed ☐ / swallowed ☐ in the stomach.
5. The last stage of digestion happens in the small ☐ / the large ☐ intestine.

Get Hands-on Experience...

Name a few things that should be chewed well while eating and also write the benefits of chewing food well.

Hands-on Learning...

Put the following food-items in their right categories :

rice	ghee	fish	papaya	sugar	oil	egg	cabbage
honey	butter	apple	pulses	nuts	sweets	spinach	meat

Proteins	Carbohydrates	Fats	Vitamins and Minerals

4 Keeping Safe

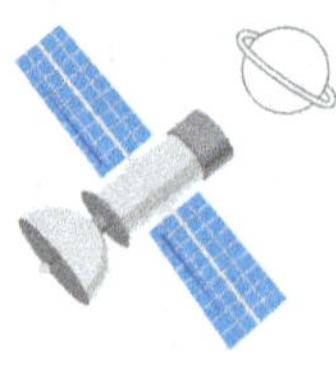

We Will Learn

- Safety at Home
- Safety on the Road
- Safety at School
- First Aid

Rack Your Brain

Sonu and Golu read in the same school. One morning, they were going to school. When they were about to cross the road while the traffic sign was red for the pedestrian crossing, Golu dragged Sonu back to the pavement. "Just think, what will happen if you cross the road without looking at signals? There is a chance of an accident, such as being struck by a speeding vehicle. We need to follow safety rules at every place to keep ourselves and others safe." So, let us learn various safety rules at different places.

Safety At Home

Falls

Falls are the most common kind of accidents at home. To avoid such accidents, always keep the floors of your house clean and dry. Take special care in the washroom.

Alertness avoids accidents

- Always go up stairs or down stairs carefully. Never rush; you may fall down.
- Do not scatter your toys, shoes or papers on the floor.
- Use a stool or ladder to reach something kept at a height.

Electric Shock

Many fires are the result of faulty electric wirings. If electrical appliances such as irons, toasters, kettles, etc. are not handled carefully, they may prove to be dangerous.

They can cause a severe electric shock or cause fire in the house. Take the following measures to avoid electric shocks.

Turn off switches before touching heaters, lamps, etc. If you see any sparks, turn off the main switch immediately.

Never touch electric wires or appliances with wet hands.

Fire

Accidents by fire mostly take place in the kitchen and can cause serious injuries and sometimes even death.

We should take proper measures to avoid such accidents.

- Do not wear nylon or synthetic clothes when near the fire.
- Never forget to turn off the gas stove when not in use.
- When gas leak is suspected, don't light matchsticks or turn on electric switches; windows should be opened immediately to let the gas out.

Poisoning

Poisoning is a common kind of an accident at home. It can be avoided by following the measures given below:

- Children should stay away from detergents, phenyl, dyes, etc.
- Medicines should not be taken at your own. The expiry dates should also be checked before taking medicines. The expired medicines may be harmful.
- Stale, spoilt and exposed food can cause food poisoning.

Safety on the Road

- Always walk on the pavement or footpath.
- Cross the road at the zebra crossing. If there is no zebra crossing, look to the right, then to the left and then to the right again. Cross only when the road is clear.

- Never play on the road.
- Look for vehicles about to take a turn at the crossroads before crossing a road.

Safety at School

- Never jump off a see-saw without informing your partner.
- Do not play rough games.
- Do not play near hedges and barbed wires.
- Do not carry pointed objects to school.
- Keep away from swings when they are in use.
- Follow the rules of the game.

First Aid

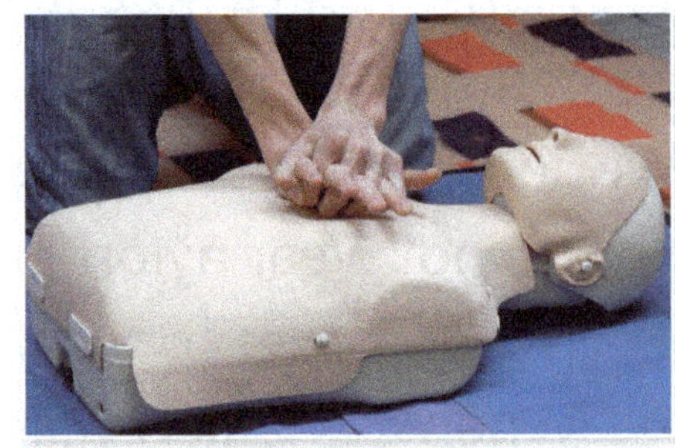

First aid is the life-saving, critical help given to an injured or a sick person before medical aid arrives. Proper and immediate first aid can save a life.

But it is very important to stay calm, not to panic, and offer help to the injured person. Each type of injury, wound or cut requires a different first-aid treatment. Some of the common methods of first aid are given below.

Take the help of your school nurse or your parents to demonstrate these aids for the first time.

Brainy Point

We should always obey the traffic lights. The red light says STOP. The amber light says WAIT. The green light says GO.

Cuts and Wounds

Wash the wound with water or any antiseptic lotion. Press the bleeding part with a piece of clean cotton cloth or cloth pad to stop the blood.

Tie a tight bandage over the pad. In case bleeding does not stop, take the injured person to a doctor.

Fainting

If a person faints, make him lie down with his head low. This makes extra blood reach his brain. Let him rest quickly.

Burns

Fire is very useful to us. But sometimes it causes burns. A burn needs special and immediate attention.

- Put ice cubes on the burn.
- Make the victim lie down and give him/her plenty of liquid to drink.
- Cover the burn with a piece of clean cloth; wash the burn with cold water.
- Take the victim to a doctor immediately.
- If a person's clothes catch fire, wrap him tightly in a blanket. This will extinguish the fire at once.

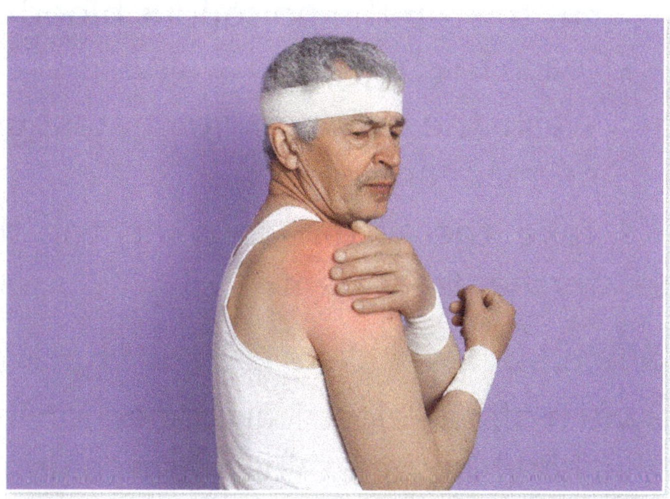

Animal Bites

The bites of some animals like dogs, cats and monkeys can be very serious. They may cause a deadly diseases called rabies. In case of an animal-bite, do the following :

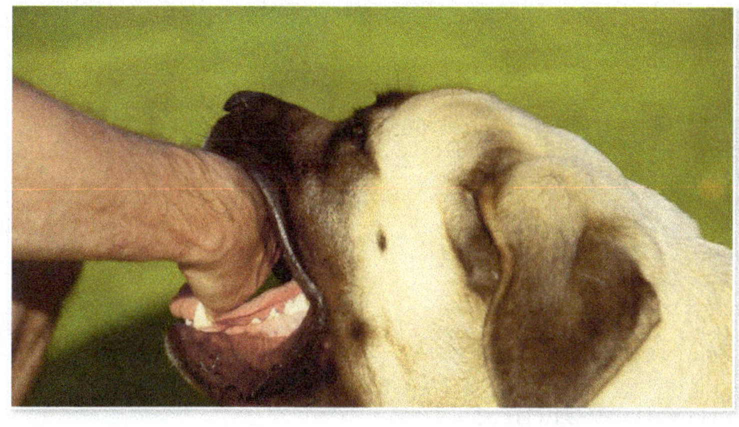

- Wash the wound with the clean water many times to remove the germs.
- Apply an antiseptic cream to prevent infection.
- Tie a bandage over the wound and take the victim to a doctor immediately.

Insect Bites

Wash the affected area with lime water and then apply a mixture of baking soda and cold cream.

Place an ice-pack over the affected area.

In case of itching, apply calamine lotion.

Always keep the first-aid box with essential items ready at home in case of an emergency.

Snake Bites

A snake-bite is very dangerous. When a snake bites, it injects its poison (venom) into the victim's blood. The poisonous blood quickly goes to the heart and brain. It stops the working of the heart and brain, and the victim ultimately dies. In case of a snake-bite, do the following :

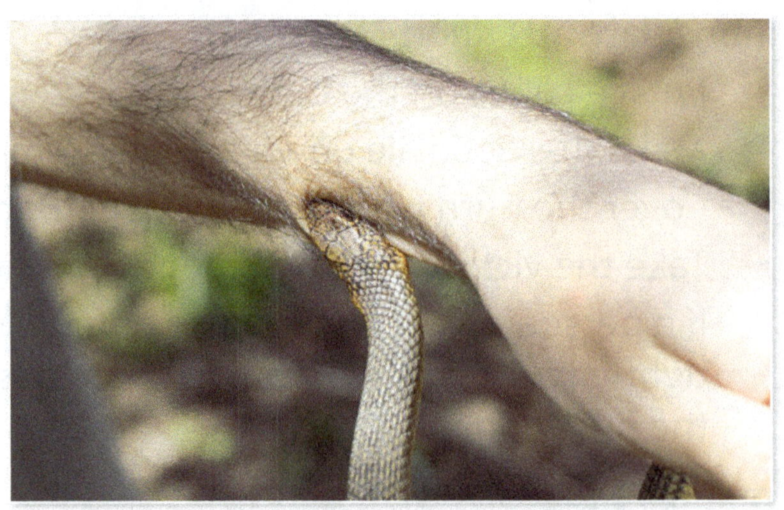

- Put a cross (×) deep over the bite and suck the blood to remove the poison.
- Tie a bandage tightly above the cut to stop the poisonous blood to reach the heart and brain.
- Wash the wound with potassium permanganate.
- Take the victim to a doctor for further treatment.

When Someone Gets Electric Shock

Switch off the electrical connection immediately. Keep the person's body warm and give him/her something warm to drink.

Summary of the Chapter

- Accidents can be prevented by following safety rules at home, schools and public places.
- At home do not run on a wet floor or in a room with furniture.
- At school do not run on stairs or slide on the banister.
- First aid can save a person's life.
- If a person's clothes catch fire, wrap him tightly in a blanket.

A. Answer the following questions :

1. What should be done to avoid accidents?
2. Why should we check the expiry date of the medicine?
3. What do you mean by first aid?
4. What would you do if your friend fainted?
5. What should we do when an insect bites someone?

B. Fill up the blanks :

1. Most accidents are caused by ______________.
2. Do not scatter your ____________, ____________ or ____________ on the floor.
3. Never ______________ on the road.
4. Follow the ______________ of the game.
5. In case of itching, apply ______________ lotion.

C. Write two safety rules that should be followed :

At Home – ____

____

On the Road – ____

____

At School – ____

____

D. Write (T) for a true statement and (F) for a false one :

1. We should stand on the kitchen shelf to get the things at high places. ☐
2. Whenever someone feels sick, he should take the medicine. ☐
3. Always walk on the pavement. ☐
4. A tight bandage should be tied over the pad. ☐
5. Always keep the first-aid box at home. ☐

Get Hands-on Experience...

Suppose you are playing in a garden. You see a boy lying on the floor unconscious. What will you do to help him?

Hands-on Learning...

A. Make a first-aid kit for your school with the help of your friends and discuss what might go into it :

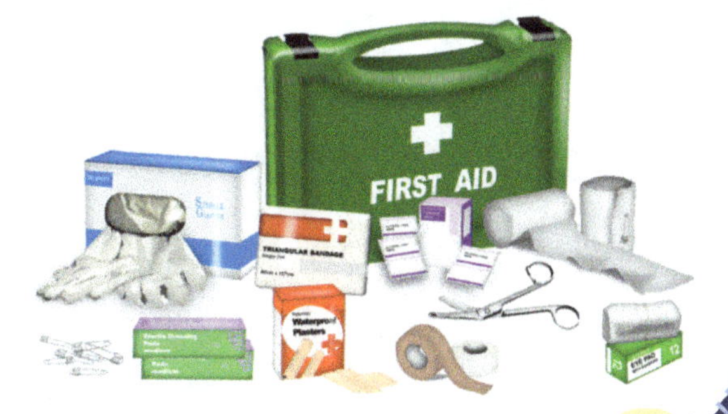

B. Write the following words under the correct signs :

No left Turn	School Ahead	Railway Crossing
No Horn	Zebra Crossing	No Parking

5 Animal Adaptations

We Will Learn

- Animal Habitats
- Animal Adaptations
- Adaptations for Living Places
- Adaptation for Food
- Adaptation for Protection
- Endangered Animals
- Extinct Animals

Rack Your Brain

Rama and her mother went to see the movie Jurassic Park. Rama was very happy to see dinosaurs. She asked her mother, "Why do we not see dinosaurs today?" Her mother replied, "Millions of years ago, dinosaurs lived on the earth. When it became very cold on the earth, they were not able to adjust to the cold climate and became extinct. Today, many animals are under threat due to the destruction of forests by humans for their needs. If we really want to protect animals in their natural habitats so that our future generations may get to see all animals that live on our planet, it is right time we changed our selfish behaviour and tried to preserve the natural habitats of animals.

Animal Habitats

A natural place where an animal lives and grows naturally is called its habitat. Animals live in different habitats such as land, water or air. Like plants, they also adjust themselves to their surroundings.

Animal Adaptations

Animals living in different habitats adapt or change themselves to suit their surroundings.

These changes occur over a long period of time. The process, in which an animal or a plant changes itself to suit its surroundings, is called adaptation. Adaptation is an essential feature of all living beings.

Most animals adapt either to their habitats, food habits or for their protection. Let us discuss adaptations in different animals.

Adaptations for Living Places

Animals live in hot deserts, cold mountains, plains, trees and water. The place where they live is called their habitat. We can make five groups according to their habitats.

Terrestrial Animals

The animals that live on the ground all time are called terrestrial animals. They have large legs to walk or run. Animals like snakes do not have legs to walk or run; they have scales to help them crawl. They have lungs to breathe air and well-developed sense organs to find food and sense safety or danger.

Rabbit Goat

Camel

Lion Horse

Animals living in cold places like sheep, rabbits and polar bears have thick fur to keep them warm. Desert animals have thick skins to protect them from the Sun. Due to the scarcity of food during cold periods animals like bears, frogs, snakes and lizards prefer to sleep throughout winter and remain inactive. This long winter sleep is called hibernation.

Camels have broad feet with thick pads. These spread out as the camel walks. Broad feet and pads make it easy for the camel to walk on the desert sand.

Brainy Point

A camel drinks a large amount of water at a time and, therefore, it is able to stay without water for long periods of time. The hump of the camel contains a lot of stored fat.

Aquatic Animals

Animals living in water are called aquatic animals. Fish, turtles and crabs are aquatic animals.

Fish swim with the help of fins. Turtles push water with their paddle-like limbs or flippers to swim. Fish and crabs have gills to breathe in water. Turtles and crocodiles have lungs to breathe.

They take out their noses above water and breathe air.

Flippers of Turtle

Crab

Fish

Aquatic Animals

Amphibians

These animals can live both on land and in water. They have limbs that are suited to help them to move both on land and in water. They can breathe with the help of lungs and also through their moist skins.

Toad

Newt

Salamander

Amphibians

Arboreal Animals

These animals live on the branches of trees most of their time. They have powerful legs and arms. Monkeys, tree lizards, flying squirrels, opossums, etc. are arboreal animals. Monkeys use their tails to grip branches and hang down for swinging. Sharp claws help squirrels and lizards to climb up trees.

Brainy Point

Koalas have become endangered because they eat only the leaves of the eucalyptus tree which has become less in number.

Tree lizard

Flying squirrel

Ape

Arboreal Animals

Aerial Animals

Birds, bats and some insects are aerial animals. These animals spend a lot of time in the air. So, they have wings to fly.

Their bones are light and hollow. Their boat-shaped bodies help them to push through the air. Many birds have such toes and claws that help them to perch on the branches of trees safely.

Parrot

Kingfisher

Bat

Aerial Animals

Adaptations for Food

We know that all animals do not eat the same type of food. Some eat only plants, some eat flesh and the rest eat both plants and flesh. The mouth parts and limbs of animals are also adapted to the type of food they eat.

Herbivores : Plant-eating animals or herbivores have straight edged teeth in the front to bite grass and broad teeth to grind leaves to a fine paste. For example– cow, ox, goat, etc.

Carnivores : Flesh-eating animals or carnivores have sharp tearing teeth and strong chewing teeth to tear flesh and chew bones. They also have sharp claws to catch their prey. For example – lions, tigers, cats, etc.

Omnivores : Animals such as bears, crows and human beings eat both plants and the flesh of other animals. They are called omnivores.

Scavengers : These animals eat the remains of dead animals. This helps to keep the environment and forest area clean. For example – vulture, hyena, hawk, etc.

Parasites : Parasites are the tiny animals that live on the blood of other animals. Lice, which are sometimes found in the hair on our heads, are parasites. Wingless insects like fleas and ticks live on birds and animals. The animal on which the parasite depends for nutrition is called the host.

Herbivore

Carnivore

Omnivore

Scavenger

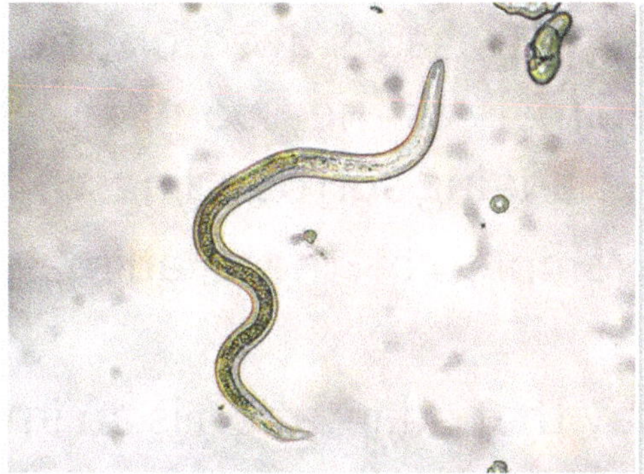

Parasite

Adaptations for Protection

1. Some animals have a great ability to run or fly to protect themselves from their enemies.
2. Some insects have hard outer body coverings for protection, such as cockroaches.
3. The hard shells of crabs and turtles protect their soft bodies from getting damaged and from danger.
4. Camouflage is the resemblance of animals to their natural surroundings giving some protection from enemies. Some animals can hide because of their features of mixing or matching with their surroundings or their colours.

5. The sharp spines of porcupines protect them from their enemies.

Chameleon

Polar bears and foxes have white fur. They cannot be seen easily against the white snow.

Tigers have black stripes over orange-coloured bodies which help them to blend with the grassland.

Colour changing is shown by some animals like chameleons and frogs. These animals hide in the surroundings by changing their colours accordingly.

Endangered Animals

Tiger

Many animals are under threat due to the destruction of forests by humans for their needs. Animals such as tigers, cheetahs, giant pandas and Indian rhinoceroses are fast declining and they may soon disappear. The animals that are in danger of disappearing from the earth are called endangered animals.

Extinct Animals

Many years ago, dinosaurs lived on the earth. Today, they have died out. We call them extinct animals as they do not exist now. They died because they could not adjust to their changing surroundings.

Some more examples of these animals are dod, Caribbean monk seals, tecopa pupfish, etc.

Thus, we can say that adaptation is an important factor for survival.

Summary of the Chapter

- Habitat is the natural home or natural environment of living things.
- According to food habits, animals are grouped into herbivores, carnivores, omnivores and parasites.
- If an animal is not adapted to its environment, it will not survive and may become endangered or extinct.
- Many animals are under threat due to the destruction of forests by humans for their needs.

A. Answer the following questions :

1. What do you understand by adaptation?
2. How are claws useful to arboreal animals?
3. How does a snake adapt to move without legs?
4. What are the ways by which animals protect themselves?
5. What are parasites?
6. What is hibernation?
7. What is a host?

B. Fill up the blanks :

1. A natural place where an animal lives and grows naturally is called its ____________.
2. ____________ is an essential factor of all living beings.
3. We can make ____________ groups of animals according to their habitats.
4. ____________ and ____________ make a camel easy to walk on sand.
5. Fish and crabs have ____________ to breathe in water.

C. Give reasons :

1. A polar bear has white fur so that ________________________
2. A cat has pointed tearing teeth ________________________
3. Dinosaurs do not exist now because ________________________
4. Tigers are in the danger of disappearing from the earth ________________________

D. Define the following :

1. Omnivores ____________________
2. Carnivores ____________________
3. Herbivores ____________________
4. Scavengers ____________________
5. Parasites ____________________

Get Hands-on Experience...

Organize a trip to your nearby sanctuary to watch the beauty of wildlife sanctuaries and national parks in India. They help you understand animals, natural habitats and survival techniques.

Discuss the trip with your friends, and form a club to protect wildlife. Name the wildlife sanctuary near your city.

Hands-on Learning...

Collect some pictures of terrestrial, aquatic, aerial and arboreal animals from old newspapers and magazines. Paste these pictures in a scrapbook and show it to your class teacher.

6 Reproduction in Animals

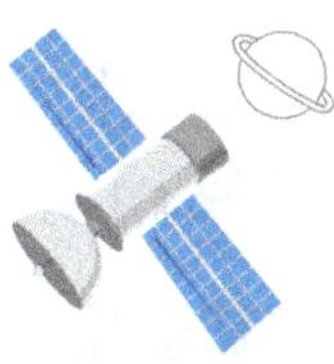

We Will Learn

- Reproduction
- Animals that Give Birth to Babies
- Egg-Laying Animals
- Life Cycle of a Bird/Hen
- Structure of an Egg
- Life Cycle of Reptiles
- Life Cycle of Insects

Rack Your Brain

Sonu and Monu are good friends. One day when they were playing in the garden, they saw the eggs of a butterfly hatching into a caterpillar. They were confused. They went to the teacher and asked about this. The teacher replied, "A butterfly lays eggs on the leaves of a plant. After a few days, the eggs hatch into caterpillars."

Similarly, all animals have the ability to produce their young ones. That is why we get to see little puppies, baby rabbits, cute kittens, tiny chicks and the calf with its mother cow. In this chapter, we will discuss the different methods of reproduction in animals .

Reproduction

An animal needs to produce one of its own kind; otherwise, it would die out.

This ability to produce young ones of its own kind is called reproduction.

Thus, reproduction is a process by which living things produce their young ones.

However, all animals do not produce in the same way.

Some lay eggs that hatch into young ones.

Some directly give birth to their young ones.

Animals That Give Birth to Babies

The animals that give birth to their babies are called mammals. Puppies, kittens, calves, colts, kids, human beings' children are born in this way. Babies feed on their mothers' milk till they are able to find food for themselves. Mammals have well-developed brains. They have hair on their bodies and most of them live on land. They breathe through their lungs. Dogs, cats, cows, horses, goats and deer are mammals. We human beings are also mammals.

Monkey

Deer

Rabbit

Some mammals such as dolphins and whales live in water. They are called aquatic mammals. They do not have hair on their bodies. They breathe through lungs and often come up to the water surface to breathe in air.

Dolphin

Whale

Egg-Laying Animals

Some animals lay eggs. Birds, insects, frogs, snakes and most of the fish reproduce by laying eggs. These animals are called egg-laying animals.

Fish

Snake

Frog

Bird

Eggs of Different Animals

Life Cycle of Bird/Hen

Female birds lay eggs in the nest. The baby bird grows inside the egg. The baby bird gets food from the yolk in the egg. The yolk supplies food to the growing bird.

While the baby bird grows inside an egg, the egg needs to be kept warm.

The father or mother bird sits on these eggs to keep them warm. This is known as incubation. When the baby birds are fully developed, they break the eggshells and come out. This is called hatching.

When the baby birds come out , they cannot fly as they are featherless and cannot see. Their parents help them feed and take care of them until they develop feathers and learn to fly.

Incubation

Hatching

Chicks

Structure of an Egg

The bird's eggs are made of hard outer shells. Within the shell there is a white jelly-like substance, rich in protein called the albumen. In the middle of the egg is yolk which is rich in vitamins and minerals. The chick forms and grows inside the yolk. The growing chick is called an embryo.

The embryo uses the yolk as its food. It develops into a chick only when the egg is kept warm.

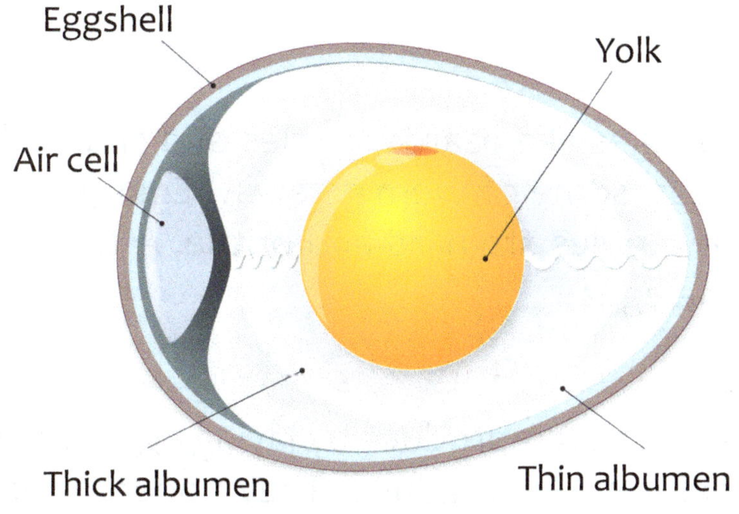

Structure of an egg

Fish

Fish lay thousands of eggs at a time. The eggs of fish do not have shells. They are surrounded by a kind of jelly. Many eggs and young fish are eaten by other bigger fish.

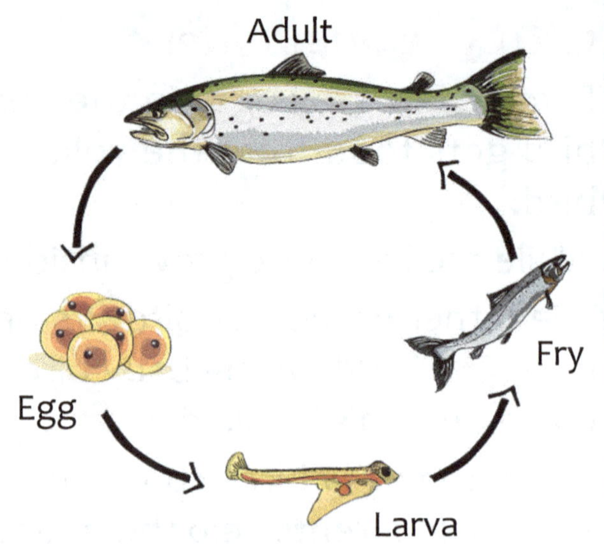

Life cycle of a fish

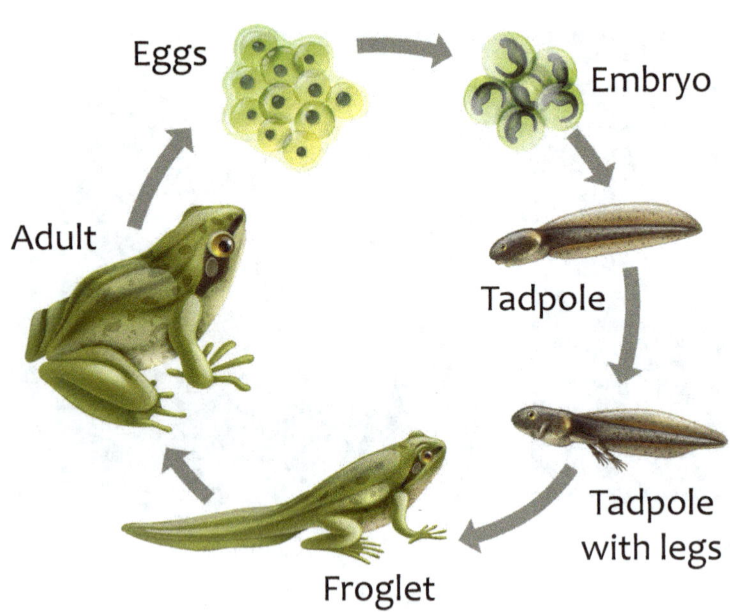

Life cycle of a frog

Frog

Frogs spend a lot of time on land. But they lay their eggs in water. The life cycle of a frog has four stages.

The egg cluster of the frog, called spawn, develops into a tadpole. Then it grows and slowly changes Into a young frog; the legs grow longer and the tail grows shorter. Then it grows into an adult frog, which has no tail. The process of growing into an adult frog from the tadpole is called metamorphosis.

Life Cycle of Reptiles

Reptiles like lizards, turtles and snakes lay their eggs on the ground. After laying eggs, some of them go away. Eggs get warmth from the heat of the Sun. The eggs of reptiles are hard and brittle. After a few days, the baby reptiles hatch out from the eggs.

Life Cycle of Insects

All insects reproduce by laying eggs. The young ones of most of the insects are very different from adults. They undergo many changes before becoming adults. This process is called metamorphosis.

Butterfly

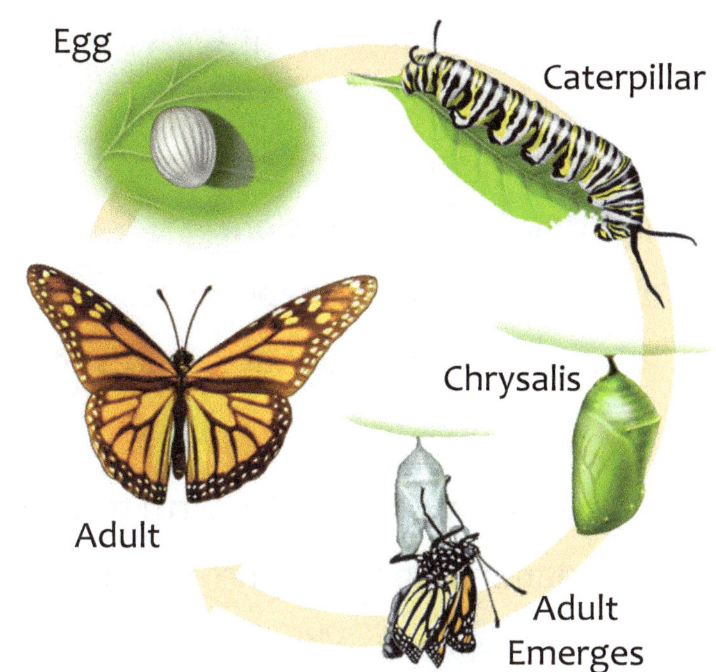

Life cycle of a butterfly

Insects like butterfly undergo four stages of growth and development. The young one that hatches from the egg is very different from the adult. It looks like a worm; it is called a larva. The larva of a butterfly is called a caterpillar. The larva of a housefly is called a maggot.

The larva feeds and grows rapidly. After some time it stops eating and makes a covering for itself. It is now called a pupa. Inside, the caterpillar keeps on changing. In a few weeks, the pupa bursts open and an adult butterfly comes out.

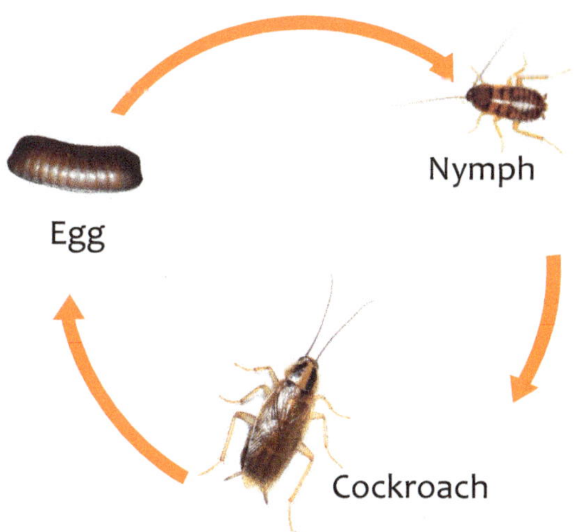

Life cycle of a cockroach

Cockroach

Insects like grasshopper and cockroach have three stages each in their life cycle. After hatching from the egg, the young one looks like the adult but it is wingless. It is called a nymph. It becomes an adult after shedding its old skin. This process is called moulting.

Summary of the Chapter

- All animals need to reproduce so that their kinds may continue to live on the earth.
- Reproduction is a process by which animals produce their young ones.
- Some animals lay eggs. Some others give birth to young ones.
- The life cycle of an animal involves the stages of the development of an embryo into an adult.
- The transformation of a young one into an adult which looks completely different is called metamorphosis.

A. Answer the following questions :

1. What is reproduction?
2. Why do animals reproduce?
3. What are the special features seen only in mammals?

B. Fill up the blanks :

1. The animals that give birth to their babies are called ______________.
2. Mammals breathe through their ______________.
3. Fish reproduce by ______________.
4. The white jelly- like substance in the egg is called ______________.
5. The young cockroach is known as a ______________.
6. The egg cluster of a frog is called a ______________.

C. Draw, label and explain :

1. The life cycle of a butterfly
2. The life cycle of a frog
3. The parts of a hen's egg.

D. Tick (✓) the correct answer :

1. Which of these is a mammal?

   a. Frog ☐ b. Cat ☐ c. Parrot ☐

2. In the life cycle of a butterfly, it comes out after the egg has hatched.

   a. Caterpillar ☐ b. Pupa ☐ c. Adult ☐

3. The animal which does not pass through the four stages of growth ______________.

   a. Housefly ☐ b. Butterfly ☐ c. Grasshopper ☐

4. ______________ take care of their young ones.

   a. Birds ☐ b. Snakes ☐ c. Frogs ☐

5. A caterpillar is the larva of a __________.

a. Mosquito ☐ b. Butterfly ☐ c. Cockroach ☐

Get Hands-on Experience...

Work in pair to create a story showing the value of 'Parents' love for their children'.

Hands-on Learning...

Draw and label the life cycle of a butterfly:

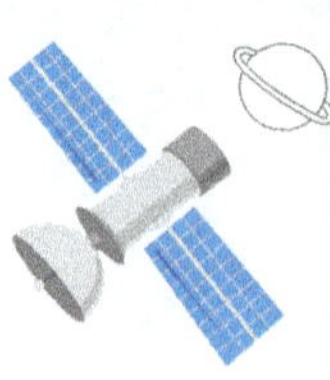

7 Plants – the Producers

We Will Learn

- The Leaf – Food Factory of the Plant
- Structure of a Leaf
- Photosynthesis
- Exchange of Gases
- How Do Plants Make Their Food?
- Interdependence between Plants and Animals
- Balance in Nature

Rack Your Brain

Dear children, every year we celebrate Van Mahotsava in our country. This promotes the planting of new trees. As we know, trees are very important to us. They provide food to eat and oxygen to breathe. They play an important role in the maintenance of balance in Nature. All of us eat fruits and vegetables every day. But have you ever thought where plants get their food from?

Green plants make their own food. So, they are called producers. In this chapter, we will learn in detail how green plants make their food.

The Leaf – Food Factory of the Plant

We use heat and many other ingredients like water, salt, etc. to cook our food. Similarly, plants also make food with the help of their green leaves.

All leaves have similar parts.

Structure of a Leaf

The flat part of a leaf is called a leaf blade. Leaves have tiny tubes called veins which bring water and minerals absorbed by the roots to the leaves and also carry back food prepared by them to the other parts of the plant.

A leaf is made up of small sections called cells. These cells contain a green-coloured pigment called chlorophyll. The chlorophyll helps to trap sunlight. There are kidney–shaped cells present on the leaf called stomata. They help the plant to breathe.

Stomata help to open and close small pores that are present on the lower side of the leaf for the exchange of gases to take place. They also help to throw out water produced during photosynthesis.

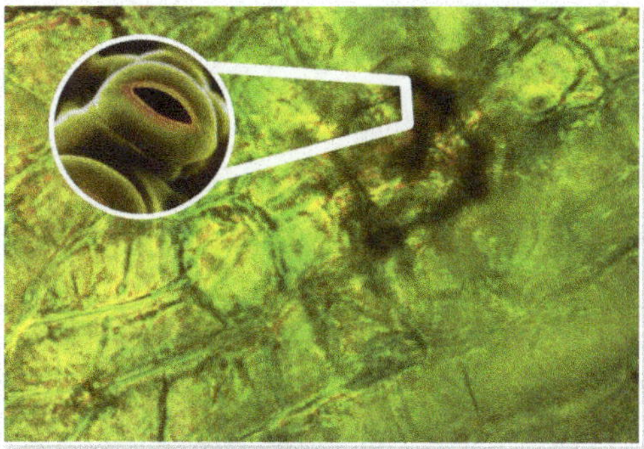

Stomata

Veins of Leaf

Apex—the tip of the leaf

Lamina or leaf blade—the broad flat green part

Midrib—the main vein running along the centre

Side vein—a number of veins found all over the leaf

Leaf stalk—this part attaches a leaf to the stem or branch

Photosynthesis

The process by which a plant prepares its food using raw materials like water, carbon dioxide and the green pigment called chlorophyll in the presence of sunlight is called photosynthesis.

Brainy Point

Moulds and mushrooms do not make their own food. They get their food from dead and decaying matter.

Since this process takes place only in sunlight, it is called photosynthesis. We can represent this recipe as follows:

water + minerals + carbon dioxide $\xrightarrow[\text{Chlorophyll}]{\text{sunlight}}$ $\xrightarrow[\text{(food of plant)}]{\text{sugar + oxygen}}$

The food is produced in the form of sugar. The plant uses this sugar to grow and live.

Exchange of Gases

All living things including plants take in oxygen and give out carbon dioxide during respiration.

However, during photosynthesis, plants give out oxygen and water vapour. The food made by the leaves is in the form of sugar.

Let us do some experiments to show the importance of chlorophyll and sunlight in the process of photosynthesis.

Experiment – I

To test that chlorophyll is essential for preparing food—

1. Take a coleus leaf. Observe it carefully. It has two colours on it.
2. Boil the leaf, first in water and then in alcohol (bleach it).
3. Wash the leaf and put a few drops of iodine on it.
4. The colour of the green part of the leaf will change into blue-black. While the non-green part will remain unchanged. This shows that food is prepared only in the presence of chlorophyll in a green leaf.

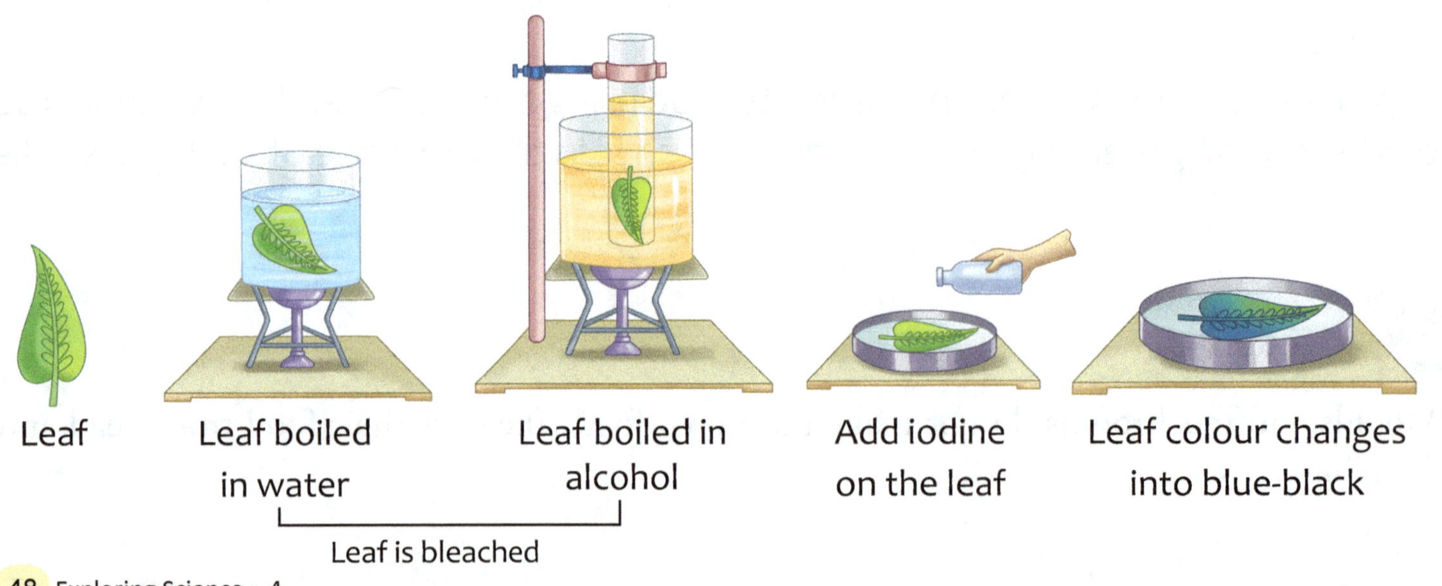

Experiment – II

To show that sunlight is needed for photosynthesis—

1. Place a potted plant in shade for 24 hours.
2. Cover one of its leaves with a black strip.
3. Now keep this potted plant again in sunlight.
4. Pluck the covered leaf. Boil it first in water and then in alcohol (bleach it).
5. Wash the leaf under water and put some drops of iodine on it.
6. The covered plant of the leaf does not turn blue-black as there is no starch in it. This shows that the process of photosynthesis takes place in sunlight only.

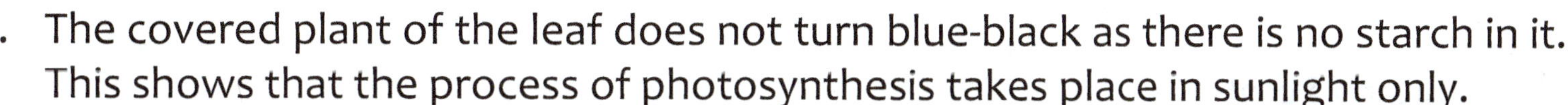

How Do Plants Make Their Food ?

The food prepared by leaves in the form of glucose or sugar is used by the plant for its growth and getting energy. The remaining food is stored in the form of starch in leaves, stems, roots or fruits. The food stored in fruits, roots, leaves and stems is eaten by us.

Interdependence Between Plants and Animals

Both plants and animals are dependent on each other in various ways for their survival. As green plants are the main food producers, they also give out oxygen. The food produced by them is eaten by human beings and human beings get energy from this. Oxygen is used for breathing by all living beings. On the other hand, animals help plants by carrying their seeds to far-off places where they can germinate. Animals breathe out carbon dioxide which is used by plants for photosynthesis.

Interdependence Between Plants and Animals

Balance in Nature

The interdependence between plants and animals creates a balance in Nature.

If the number of either plants or animals increases or decreases suddenly, this balance will be disturbed.

If there is a sudden decrease in the number of plants in Nature, the oxygen level in the atmosphere will decrease leading to an increase in the level of carbon dioxide. Similarly, if there is sudden increase in the number of animals in Nature, it will lead to an outbreak of diseases and insufficient food to meet the needs of growing population.

Brainy Point

Sugarcane and bamboo are also the plants of the grass family but instead of the grains their stems are used. A sugarcane stem can be used for making a special fuel which is similar to petrol or gasoline.

Summary of the Chapter

- Green plants are called producers.
- Photosynthesis is the process by which plants make their food from water and carbon dioxide in the presence of sunlight.
- Plants make food in the form of starch or sugar.
- Apart from food, plants also provide us with many other things like wood, paper, gum, cloth, rubber, etc.
- Plants and animals depend on each other for their survival.

A. Answer the following questions :

1. How does chlorophyll help the plant?
2. What do you understand by photosynthesis?
3. Which gases are exchanged during photosynthesis?
4. How do plants use their food?
5. How do plants and animals depend on each other?

6. Why is it necessary to maintain balance in Nature?
7. Write the experiment to show that sunlight is needed for photosynthesis.

B. Fill up the blanks :

1. The flat part of a leaf is called a ______________.
2. The food produced is in the form of ______________.
3. The ______________ is used for breathing by all living beings.
4. Animals breathe out ______________.
5. A leaf is made up of small sections called ______________.

C. Write a single word for each of these :

1. The food made by plants. ______________
2. The holes found on leaves. ______________
3. The process of food-making by plants. ______________
4. The liquid used for testing the presence of starch. ______________
5. Kitchen of the plant. ______________

D. Tick (✓) the correct answer :

1. The tiny tubes on leaves are called ______________.

   a. Cells ☐ b. Stomata ☐ c. Veins ☐

2. The green colour pigment in leaves is called ______________.

   a. Photosynthesis ☐ b. Chlorophyll ☐ c. Starch ☐

3. Plants take in ______________.

   a. oxygen ☐ b. carbon dioxide ☐ c. air ☐

4. The solution used for the chlorophyll test is ______________.

   a. Iodine ☐ b. glucose ☐ c. sugar ☐

Get Hands-on Experience...

Visit a nursery in your city or town. Find out the names of any four plants grown there and their uses.

Hands-on Learning...

Look at the pictures given below and match them with the messages they are giving:

1. Do not walk on the grass with shoes on.

2. Never pluck leaves.

3. Each one grows a plant.

4. Never cut trees.

8 Plant Adaptations

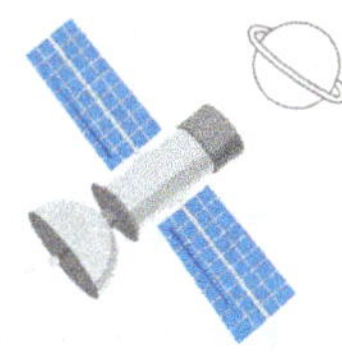

We Will Learn

- Plant Habitats
- Terrestrial Plants
- Aquatic Plants
- Parasitic Plants
- Insectivorous Plants

Rack Your Brain

Shyam and his father were walking in a garden. Suddenly Shyam said, "Dad, we find plants everywhere around us, whether it is high mountain tops, sea coasts, deserts or crowded cities. But how can they survive in such varied environments?"

Shyam's father said, "It is very important to change according to the environment. Many plants and animals have become extinct because they did not change." In this chapter, we will learn about the ability of plants to change themselves according to the environment in order to survive.

Plant Habitats

Plants are found all over the world. They live and grow in different surroundings. Some plants grow in water whereas others grow on land. Some plants grow at cold places whereas others grow in hot deserts. These regions are called their habitats.

Plants growing in a particular habitat develop some specific features that help them to survive and flourish there. This process is called adaptation.

On the basis of habitats all plants can be broadly divided into two types:

1. Terrestrial Plants
2. Aquatic Plants

Terrestrial Plants

Most of the plants grow on land. These are known as terrestrial plants.

Terrestrial plants can further be classified as follows:

Plants in Hilly Regions

Features of these plants

These are generally tall and thin. This helps them trap sunlight.

These trees are generally cone-shaped with sloping branches. It helps to slide off snow easily. The leaves of such plants are needle-like. This helps leaves to face cold and snow. They do not bear flowers.

They bear cones. Their seeds grow inside cones.

The plants of hilly regions are known as conifers because of their shape.

Fir

Pine

Deodar

Spruce

Plants in Hilly Regions

Brainy Point

There are about 600 types of carnivorous plants found all over the world except in Antarctica.

Plants in Plain Regions

The trees that grow in the plains have the following features–

1. These trees have lots of leaves with branches spread around.
2. Their leaves are flat and thin that help them trap a lot of sunlight to make food.
3. These plants or trees shed all their leaves mostly in winters.
4. When spring comes, new leaves grow again.

Mulberry

Peepal

Sheesham

Neem

Plants in Plain Regions

Plants in Deserts

The plants of deserts develop some special features which help them to survive in deserts without water for many days.

The leaves of these plants are reduced to spines. This feature helps to reduce water-loss from plants. The process of photosynthesis in these plants takes place in the stem instead of leaves. The fleshy stem stores food and water which is used by the plant to survive in deserts.

Desert plants have very long roots. Their roots spread deep into ground in search of water. Some plants have strong and hardy leaves. This helps them to survive in hot and dry deserts.

Cactus

Date Palm

Plants in Deserts

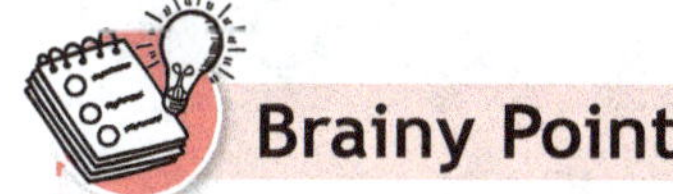

Brainy Point

Saguaro is a type of cactus found in deserts. It can live up to 200 years and can store up to 800 litres of water.

Plants in Marshy Areas

The trees that grow in marshy plains are called mangroves. The roots of mangroves grow above the soil because air cannot reach their roots. This type of root is called breathing root. These roots have tiny openings at their tips, through which air can pass.

Mangroves

Plants in Heavy Rainfall Areas

Some smaller plants that grow in heavy rainfall areas are full of leaves and remain green for the whole year. These plants are called evergreen trees.

Teak

Sugarcane

Cotton

Plants of Heavy Rainfall Areas

Plants in Coastal Areas

The plants that grow in coastal areas are adapted to hot and humid climate. These trees also do not shed their leaves in winter, so these plants are also known as evergreen trees.

Coconut trees grow in coastal areas because they survive in salty water.

Rubber

Coconut

Pepper

Plants in Coastal Areas

Brainy Point

Some of the tallest plants in the world are the Redwood trees that grow to a height of about 300 feet.

Aquatic Plants

The plants that grow in water are called aquatic plants. The first plants on the earth were the water plants over 3 billion years ago. Aquatic plants are of three types:

- Floating Plants
- Fixed Plants
- Underwater Plants

Floating Plants

The aquatic plants which float on the surface of water are called floating plants. Their roots are not fixed in the soil at the bottom of water. They float because they are small, light and spongy.

Water Hyacinth

Duckweed

Water Chestnut

Fixed Plants

These plants remain fixed in the soil at the bottom of a pond by long hollow stems and deep roots. Their broad leaves float on the surface of water. Leaf stalks are spongy and filled with air. Leaves have tiny pores only on their upper surface for respiration.

Water lily

Lotus

Underwater Plants

These are fixed water plants. These plants remain completely immersed in water. These plants have narrow ribbon-like leaves but no pores. They breathe dissolved air by their entire body surface and this helps them keep the water clean.

Underwater plants are also grown in an aquarium in order to keep it clean.

Pondweed

Tapegrass

Hydrilla

Parasitic Plants

Plants like mushrooms and moulds do not have chlorophyll. Hence, they cannot make their own food. They absorb food from other plants or decaying matter. They are called parasitic plants.

Mushroom

Mould

Insectivorous Plants

The plants that trap and eat small insects are called insectivorous plants. Whenever an insect sits on the leaf of this plant, the two folding halves of the leaf get folded, press and absorb the insect.

These plants are usually found in the areas where there are not enough nutrients in the soil for healthy growth.

Summary of the Chapter

- Plants adapt in order to survive.
- Aquatic plants adapt themselves by developing special features to float or stay completely under the water.
- Mangroves grow in marshy areas. These plants have special breathing roots that come out of the soil.
- Insectivorous plants adapt their feeding habits in order to survive.
- Plants cannot survive if their habitat is changed.

A. Answer the following questions :

1. On what basis can different plants be classified?
2. What is adaptation?
3. What are the special features of fixed aquatic plants?
4. How does a cactus survive in deserts?
5. How does a venus flytrap plant trap insects?
6. Give two features of the plants of terrestrial plants.

B. Tick (✓) the correct answer :

1. Terrestrial plants are the plants which grow ________________.
   a. on land ☐
   b. in water ☐
   c. under water ☐
2. Venus flytrap is ________________.
   a. a decaying plant ☐
   b. an insectivorous plant ☐
   c. a non-green plant ☐

3. Duckweed is ______________.

   a. a fixed plant ☐

   b. a floating plant ☐

   c. an underwater plant ☐

C. Fill up the blanks :

1. Cone-shaped leaves help ______________ to slide off easily.
2. The plants of ______________ region do not bear flowers.
3. When ______________ comes, new leaves grow again.
4. The leaves of ______________ plants are reduced to spines.
5. The plants of ______________ and ______________ areas are known as evergreen trees.
6. Plants like ______________ and ______________ cannot prepare their own food.

D. Name the following :

1. A plant with spines ______________
2. A tree with cone-shaped leaves ______________
3. An insectivorous plant ______________
4. A parasitic plant ______________
5. A plant that does not bear flowers ______________

Get Hands-on Experience...

When you visit a hill station, collect the cones and leaves of some coniferous trees. Observe how the cones and leaves of different trees differ from one another.

____

____

____

____

Hands-on Learning...

Collect and paste flowers and seeds from different trees.

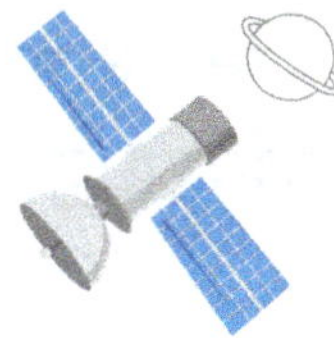

9 Air, Water and Weather

We Will Learn

- Air
- Water
- Weather

Rack Your Brain

Jaya and Pragya were walking at Marine Drive late in the morning. They could feel a noticeable change in the temperature as the cool sea breeze began to blow off the water. But as soon as they reached home, they felt very hot. Dear children! Why do you think there was a difference in the temperature? The Sun heats the land and water differently. The land heats up faster than the water and also cools faster. So, it is pleasant to visit the seashore during the day because a cool breeze blows from the sea and makes the places nearer the sea cooler than places farther from the sea. Now, let us read about them in detail.

Air

Air is a mixture of several gases and water vapour. The layer of air around the earth is known as atmosphere. Air contains 78% nitrogen, 21% oxygen and smaller amounts of carbon dioxide, hydrogen and other gases.

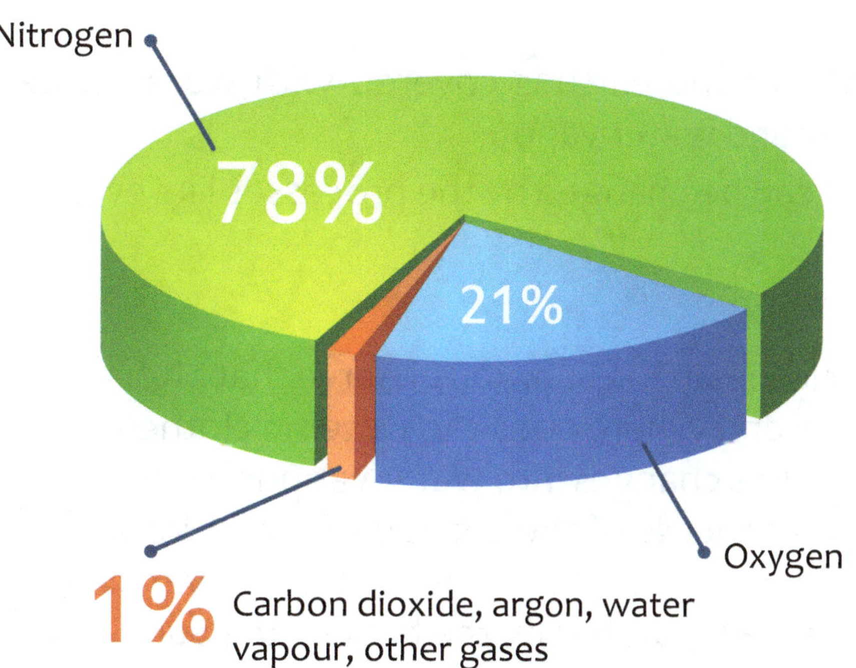

We cannot see air but feel when it blows. The moving air is called wind. Gently blowing air is called breeze whereas strongly moving air is known as gale. When the wind moves very strongly, it is called storm.

Sea Breeze

Breezes are of two types :

- Sea breeze
- Land breeze

Land Breeze

Sea Breeze

During the daytime, land gets heated faster than the sea. So, the air above the land also gets heated quickly. As the hot air is light, it rises up. So, cool air from the sea flows in to take the space of warm air. This is known as sea breeze.

Land Breeze

At night reverse process takes place. Land cools down quickly. But water in the sea is warm as it cools slowly. So, the air above the water is also hot. As the hot air is light, it rises up. The cool air from the land flows to occupy the space of warm air. This is known as land breeze.

Water

About three-fourths of the earth is covered with water. It exists in three forms, namely snow, water and water vapour.

The states of water can be changed by the processes like evaporation, condensation and freezing.

Evaporation

When you put wet clothes in the Sun, you observe that after a few hours they get dry. Where does the water in clothes go? The water of clothes changes into water vapour by the heat of the Sun. This process of the change of water into water vapour is called evaporation. The water in rivers, lakes and oceans gets evaporated by the heat of the Sun.

Condensation

You might have seen some water droplets on the lid of the pan while water is boiling or food is cooking in it. It happens due to the change of water vapour into water. This change of water vapour into water is known as condensation.

Condensation takes place in Nature too. Clouds, dew, fog and snow are its examples.

Freezing

Freezing is the process in which liquid water changes into snow.

Purification of Water

Impure water can make us fall ill. So, we should always drink pure water. Water can be purified in the following ways:

Boiling Water

Boiling

Impure water contains various harmful germs. By boiling, the water can be made germ-free. Boiling the water for 20 minutes kills germs present in it.

Sedimentation and Decantation

This process is used for purifying the water containing insoluble impurities like mud, sand, etc. In this process, water is allowed to stand in a container for a few hours. The impurities settle down at the bottom of the container. This process in which insoluble impurities settle down is called sedimentation. Then the clear water is poured off gently into another container. This process is known as decantation.

Filtration

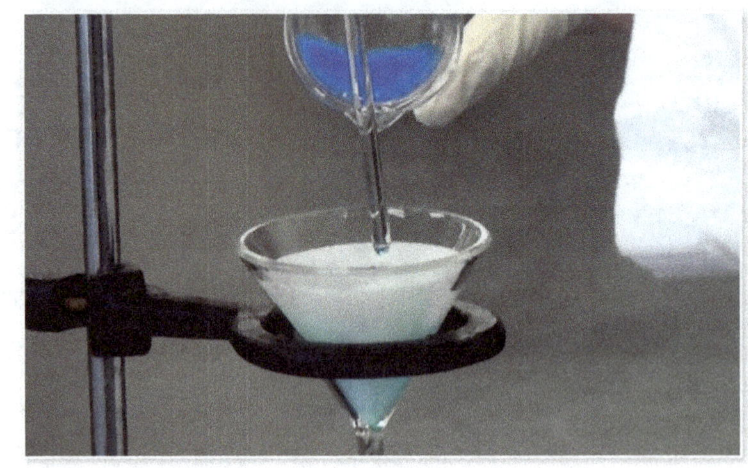

Filtration is the most common process of purifying the water. In this process, water is passed through a filter paper. Impurities are left behind on the filter paper and clean water is collected in a container below. This process of removing impurities by passing water through the filter paper is called filtration.

Chlorination

Chlorine is added to water to kill germs. This process of adding chlorine to water is known as chlorination.

Weather

The conditions of atmosphere around us keep on changing. Sometimes it is very hot while sometimes very cold. Sometimes we see heavy rain. The change of the condition of atmosphere around us at a particular time is called weather. The Sun and the earth play an important role in causing the change in weather.

The heat of the Sun causes air to blow. When the air gets heated by the Sun, it rises up. The cool air then comes to occupy the space of warm air.

The heat of the Sun also causes the evaporation of water in rivers, lakes and oceans which results in the formation of clouds and rain.

Cause of Change in Weather

As the earth is tilted on its axis, one half of the earth is closer to the Sun than the other half. The part closer to the Sun has longer and hotter days and thus has summer. The part away from the Sun has shorter and colder days and thus has winter.

Summary of the Chapter

- The Sun causes changes in weather.
- The distance from the sea also affects weather.
- Water becomes impure because of mud, stones and dissolved minerals.
- Boiling, sedimentation, decantation, filtration and chlorination are some of the methods used for purifying water.

A. Answer the following questions :

1. What is air?
2. Why do we need to purify water? Write two methods of the purification of water.
3. What is water?
4. What is evaporation? Explain with an example.
5. What is condensation? Explain with an example.
6. What is chlorination?
7. What is filtration?
8. What is weather? What is the role of the Sun in the change of weather?
9. What causes change in seasons? Explain.

B. Fill up the blanks :

1. The moving air is called ______________.
2. Clouds, dew and fog are the examples of ______________.
3. The part of the earth closer to the Sun has ______________.
4. Sedimentation is used for removing ______________.
5. We can kill germs in water by adding ______________.

C. Tick (✓) the correct answer :

1. Sea breeze occurs during daytime ☐ / night ☐.

2. The strongly moving air is called wind ☐ /storm ☐.
3. In condensation ☐ /evaporation ☐ , water changes into water vapour.
4. The heat of the Sun causes evaporation ☐ /condensation ☐ in rivers, lakes and oceans.

D. Give one-word answer :

1. How much per cent of oxygen is present in air? ____________
2. What is the gently moving air called? ____________
3. Name the process of the change of water into water vapour. ____________
4. Name the process used for purifying water containing insoluble impurities. ____________
5. Give an example of condensation. ____________

Get Hands-on Experience...

Imagine you run out of drinking water on your trekking camp. Think and describe a few ways by which you will be able to get safe drinking water.

Hands-on Learning...

Make a poster in the given space on how water and air are getting polluted by our modern lifestyle:

10 Our Universe

We Will Learn

- Our Universe
- Satellites
- Movements of the Earth

Rack Your Brain

Shanu and his sister sweety were watching a movie on Discovery Channel. It was all about visitors from other planets who came in their spaceship. It made them wonder about the mystery of the universe and the life there. They wondered how the universe and the stars were made. Are there more planets like ours? Dear children! Let us explore the wonderful universe.

Our Universe

Our solar system consists of the Sun, the moon, all the stars and eight major planets. The Sun is at the centre of the solar system. All the eight planets revolve around the Sun in their orbits.

The solar system contains Mercury, Venus, Earth, Mars, Jupiter, Saturn, Uranus and Neptune. In addition to planets the solar system also consists of moons, comets, asteroids, minor planets, dust and gas.

Brainy Point

The average distance of the Sun from the Earth is about 150 million kilometres. This distance is called one astronomical unit (1AU).

Mercury

Mercury is the closest planet to the Sun. Mercury is also the smallest planet . It is greyish in colour. It has no moon.

Venus

Venus is the brightest and hottest planet. It is the brightest object in the night sky after the moon. It is the second planet from the Sun. Venus is also called evening star or morning star as it is seen at dusk or dawn.

Earth

The earth is the only planet having life. It is also known as watery planet because 71% of its surface is covered with water. It orbits around the Sun once in about 365 earth days. The moon is the only satellite of the earth. The earth is also called blue planet due to the presence of plenty of water on it.

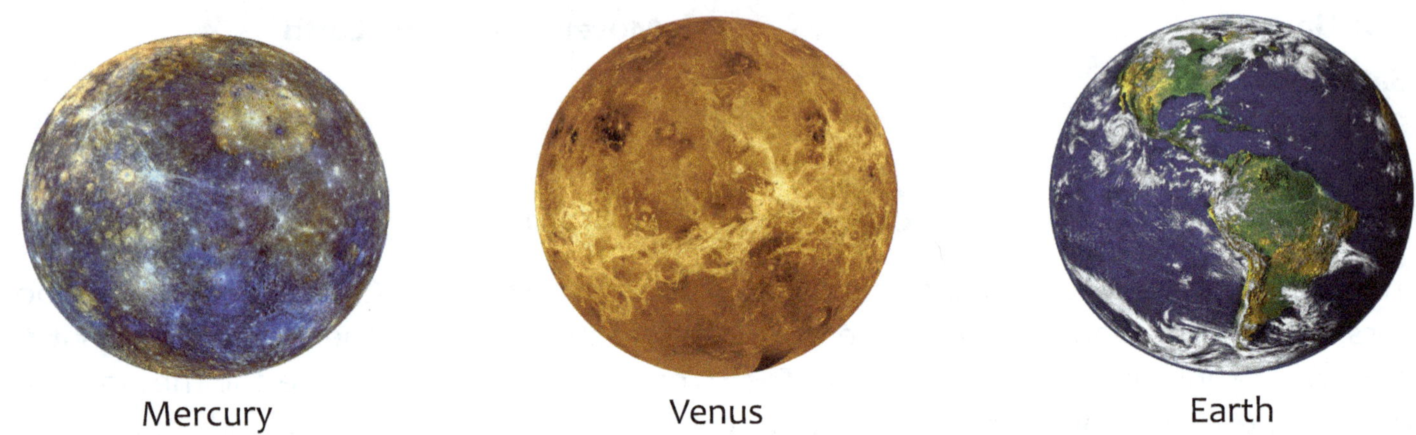

Mercury Venus Earth

Mars

Mars is the fourth planet from the Sun. It has two satellites. Mars is also called Red planet because the soil and rocks on its surface are red in colour.

Jupiter

Jupiter is the largest planet in the solar system. Jupiter rotates faster than any other planet. It completes one rotation in 10 hours. Jupiter is also the fastest spinning planet. Jupiter has about 63 moons.

Saturn

Saturn is the sixth planet from the Sun and the second largest planet in the solar system, after Jupiter. It has 62 known moons. It can be seen by the naked eye. Saturn is 95 times heavier than the earth.

Mars Jupiter Saturn

Uranus

Uranus

Uranus is the seventh planet from the Sun. It is the only planet, which rotates on its axis from east to west. Uranus has 27 known moons.

Neptune

Neptune

Neptune is the farthest planet from the Sun. Neptune's winds are the fastest winds in the solar system. It is the coldest planet. It has 13 moons. It has a great dark spot.

Brainy Point

Mercury, Venus, the Earth and Mars are the inner planets whereas Jupiter, Saturn, Venus and Neptune are the outer planets.

Satellites

An object that moves around a planet in an orbit is called its moon or satellite. Most planets have satellites. Our planet earth has only one natural satellite, the moon, whereas Jupiter has 48 moons. Mercury and venus have no satellites at all. These satellites reflect sunlight and, hence, are able to shine. There are two types of satellites:

Natural satellite

The moon is the natural satellite of the earth. It is about 3,84,400 km away from the earth. It completes one revolution around the earth in 27 days. It does not have its own light but appears bright as it reflects the light of the Sun. It does not have atmosphere, unlike the earth, to protect itself from very hot or cold conditions.

So, the temperature on the moon can rise till 117°C at noon and come down to 173°C at night. The surface of the moon is covered with large pits or holes called craters.

Artificial Satellites

The earth has a lot of man-made satellites revolving around it. These are called artificial satellites and revolve around the earth in fixed orbits. One of the first artificial satellites to be sent into space was SPUTNIK-1.

The first artificial satellite to be sent into space was Aryabhatta. It was launched on 19 March, 1975.

Stars

On a clean night, we can see lots of small twinkling objects in the sky called stars. There are millions of stars in the universe. They are found in clusters called galaxies. Smaller groups of stars making definite patterns that are found in the universe are known as constellations. So far, the constellations that have been discovered are Scorpius, Orion, Ursa Major, Ursa Minor and Hercules.

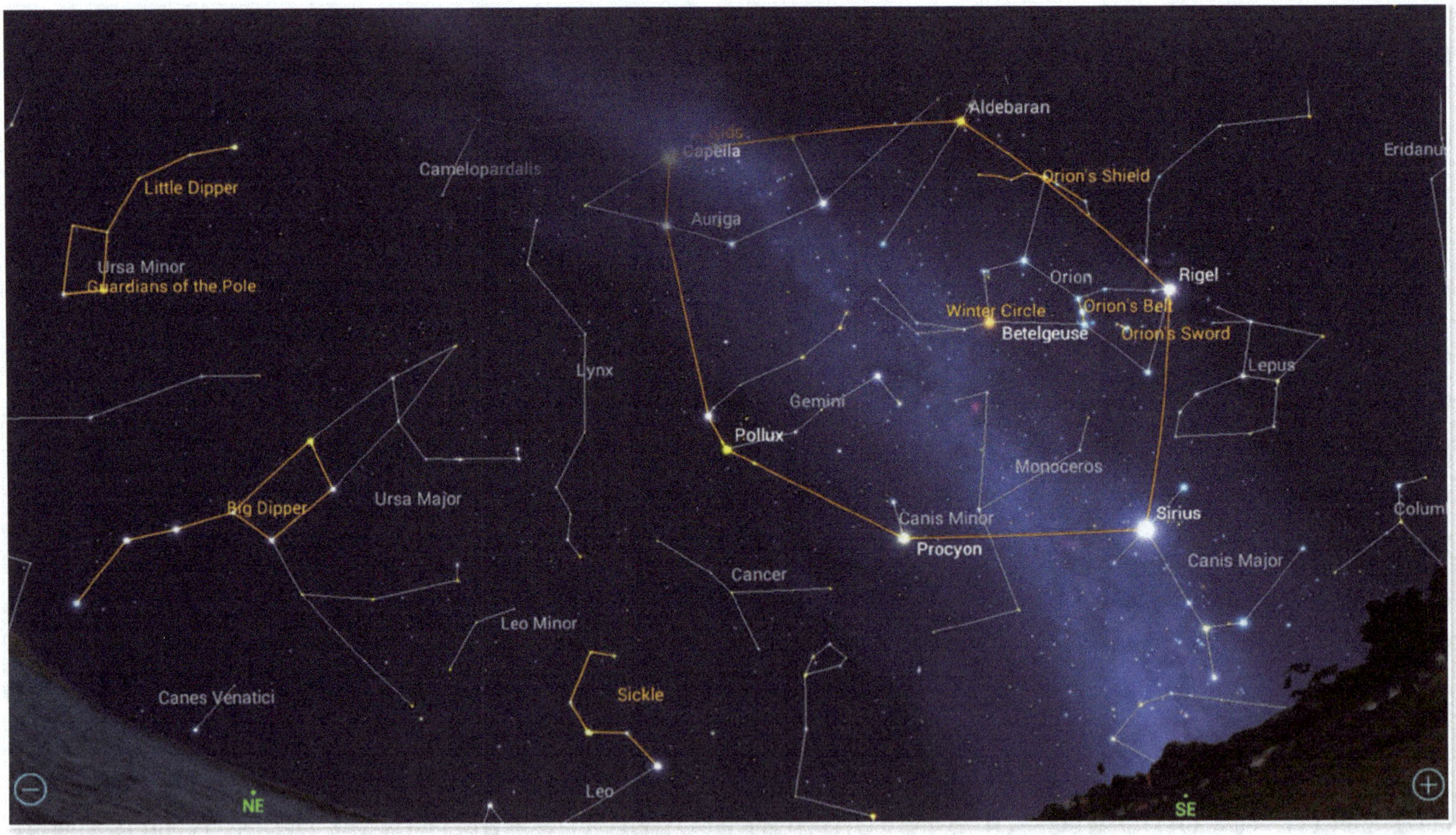

Constellations

Brainy Point

Mangalyaan is India's Mars Orbiter mission, which was successfully launched on November 5, 2013, by the Indian Space Research Organization (ISRO).

Movements of the Earth

There are two types of the movements of the earth–rotation and revolution. These movements have different effects on planet earth.

Rotation

The rotation of earth on its axis causes day and night. The axis is an imaginary line running from the North Pole to the South Pole. It is tilted at an angle of 23.5°. The earth completes one rotation on its axis in 24 hours.

You know that the earth is like a ball. When it rotates on its axis, one half of it faces the Sun, and this side has day. The other half faces away from the Sun and this side has night.

India and the United States of America are situated on the opposite sides of the earth; when we have day in India, it is night in the USA.

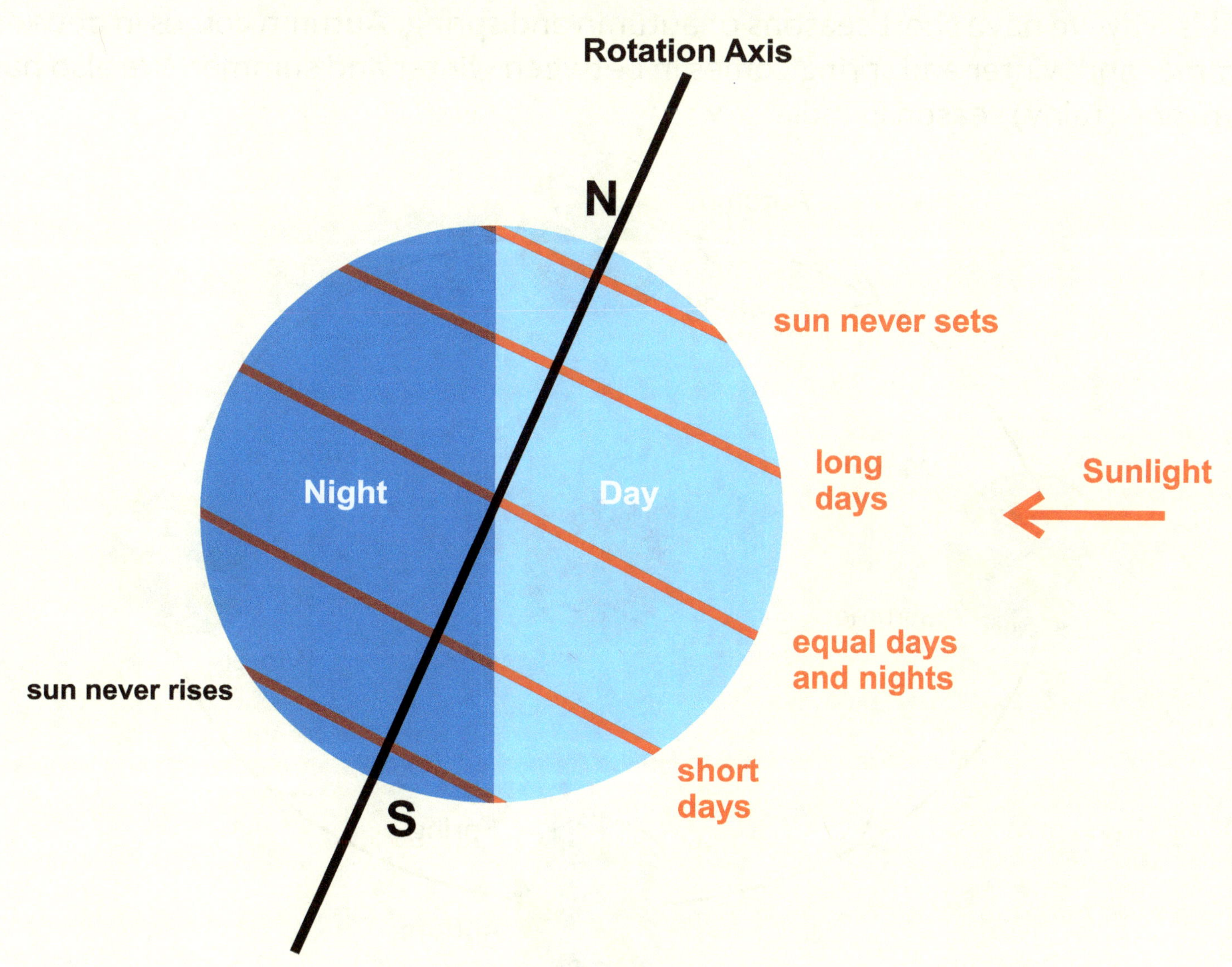

Change in Day and Night

Revolution

Revolution is the movement of the earth around the Sun.

As it moves around the Sun, one half of the earth is always closer to the Sun than the other half which is slightly away from the Sun because the earth is tilted at an angle of 23.5°. The revolution of the earth around the Sun and the tilt of the earth cause change in seasons.

There would be no seasons if the earth were not tilted on its axis. The tilt of the earth causes the Sun's rays to fall directly either on the Northern Hemisphere or on the Southern Hemisphere.

The half that receives more sunshine has summer and the other half has winter.

New Zealand lies in the Southern Hemisphere whereas India lies in the Northern Hemisphere. When New Zealand has summer, we have winter in India.

The earth does not change its position suddenly. It goes round the Sun gradually; that is why we have short seasons of autumn and spring. Autumn comes in between summer and winter and spring comes in between winter and summer. We also have monsoon (rainy) season in India.

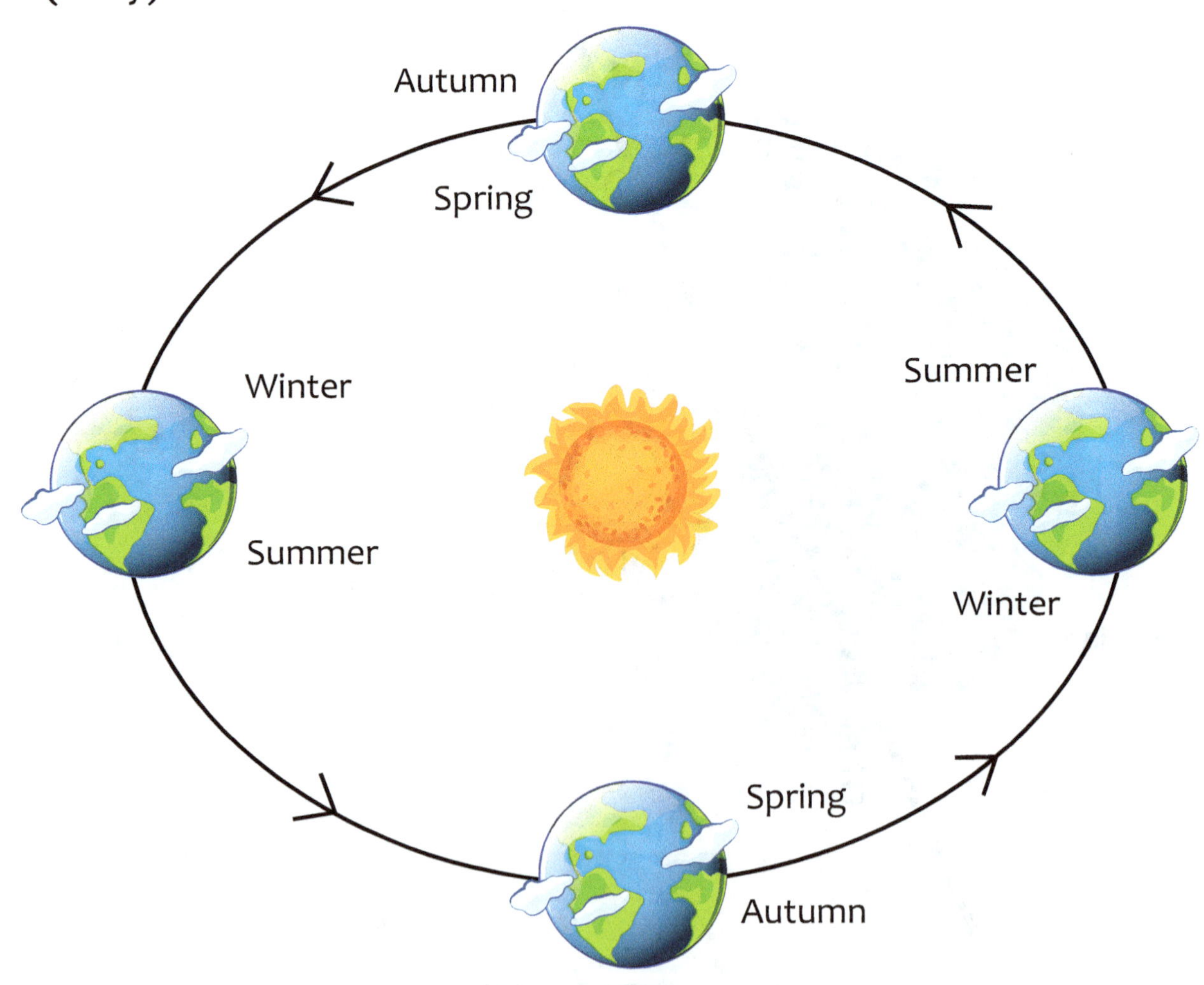

Change in Seasons

Summary of the Chapter

- The Sun with its family of eight planets and moons forms the solar system.
- The planets revolve around the Sun in their own fixed orbits.
- An object moving around a planet in the solar system is called its moon or satellite.
- Rotation and Revolution are the two movements of the Earth.

A. Answer the following questions :

1. What is the solar system? Where is it located?
2. What is an orbit?
3. Name the eight major planets of the solar system.
4. Why is the earth the only planet where life exists?
5. Which planet is also known as Red planet? Why?
6. What is the earth's rotation?
7. What is the earth's revolution?

B. Tick (✓) the correct answer :

1. Mercury/Mars is the closest planet from the Sun. ☐
2. Earth/Venus is the only planet where life exists. ☐
3. 71 %/91% of the earth's surface is covered with water. ☐
4. Uranus/Jupiter is the largest planet of the solar system. ☐
5. Venus/Saturn has rings. ☐

C. Number the planets according to their distances from the Sun :

1. Saturn ☐
2. Venus ☐
3. Jupiter ☐
4. Neptune ☐
5. Uranus ☐
6. Venus ☐
7. Mercury ☐
8. Mars ☐

D. Write one line for each of the following planets :

1. Mercury

   ____

2. Venus

   ____

3. Earth

   ____

4. Mars

   ____

5. Jupiter

   ____

6. Saturn

   ____

7. Uranus

   ____

8. Neptune

   ____

Get Hands-on Experience...

Visit a planetarium and observe the position of the Sun, the moon, stars and planets.

Hands-on Learning... 

Make a model of the solar system and display it in your classroom. Don't worry about the sizes of the planets but try to get the correct order.

11 Clothes and Fibres

We Will Learn

- We All Need Clothes
- Kinds of Clothes
- Fibres
- Care of Clothes

Rack Your Brain

Sanya had to attend a party in the evening. It was the month of December. She was wearing a woollen sweater; she felt warm and comfortable. Dear children! Can you tell why the woollen sweater kept her warm ? Because woollen clothes are thick and do not allow the heat to escape from our bodies. Similarly, during the summer season, we wear cotton clothes to keep our bodies cool. Cotton clothes absorb sweat and allow the body heat to escape. In this chapter, we will learn about clothes worn in different seasons.

We All Need Clothes

Clothes are necessary for us. We wear them because:

- Clothes keep us warm and protect our bodies from outside sources.
- Clothing is also seen as important for the presentational purposes it has.
- Clothing allows us to differentiate what role a person has in a society such as police, firefighters, etc.
- Clothes are also good for appearance and individual expressions.
- Clothing allows us to enhance our attractiveness and allows us to feel a certain way that we want to be seen as classy, casual, stylish, edgy, laid-back, etc.
- Clothes also protect us from harsh weather conditions. Woollen clothes keep our bodies warm in winters.
- We use raincoats and gumboots to keep our bodies dry in the rainy season.

Kinds of Clothes

We wear clothes depending upon two conditions like :

- Climate
- Occasion

Let us read about them in detail.

Climate

We wear clothes according to the climate of a place. We wear cotton clothes in the summer season. They keep our bodies cool and absorb the sweat from our bodies. Cotton clothes act as a barrier between the hot Sun-rays and our bodies.

In winters, we wear woollen clothes. Woollen clothes keep our bodies warm. They trap the body-heat and keep us warm. They do not let the cold affect our bodies.

Summer clothes   Winter clothes

Occasion

We wear different clothes on different occasions. The clothes which we wear at home, at school, in parties, etc. are different.

We wear our uniforms while going to school. We wear sports uniforms while playing sports. We wear night dresses while sleeping. We wear nice colourful dresses when we go for a party or any other function.

Fibres

A fibre is a thread from which clothes are made. The fibres are woven and knitted together to make a fabric. There are two types of fibres:

- Natural
- Synthetic

Let's read about them in detail.

Natural Fibres

The fibres we get from plants and animals are called natural fibres. Cotton, jute, wool and silk are natural fibres.

Cotton comes from cotton plant.

Jute comes from jute plant.

Wool comes from sheep.

Silk comes from silkworms.

Brainy Point

The cultivation of silkworms for producing silk is known as sericulture.

Nylon

Polyester

Synthetic Fibres

Synthetic fibres are man-made fibres. They are made from chemical resources. Acrylic and rayon are also synthetic fibres.

Care of Clothes

We should take good care of our clothes. We can follow these steps to make our clothes last longer and look better.

Ironed clothes

- We should wash our clothes with a good detergent and water.
- After washing clothes, we should dry them in the Sun so that the bad odour may vanish from clothes.
- We should iron our clothes. Ironing makes our clothes look neat and good.
- Washed and ironed clothes should be kept nicely in the cupboard.
- Expensive clothes made of wool and silk should be dry-cleaned. Dry cleaning saves clothes from damage.
- Woollen and silk clothes should be stored with mothballs or neem leaves. They keep insects away from these clothes.

Summary of the Chapter

- We need clothes to protect us from heat, dust, rain and cold.
- Clothes are made using materials called fabrics.
- We wear clothes that are suited to seasons.
- Regular washing and proper storing of clothes make them last longer.

A. Answer the following questions :

1. Why do we need clothes?
2. What do we wear in the summer season? Explain in detail.
3. How can we protect our bodies from insects?
4. How do we select clothes depending on the climate?
5. What types of clothes do you wear on different occasions?

6. Write the difference between natural and synthetic fibres.
7. Write two points to take care of clothes.
8. How are raincoats and gumboots useful to us?

B. Tick (✓) the correct answer :

1. We wear cotton [ ] / woollen [ ] clothes in summer.
2. Cotton [ ] / Synthetic [ ] fibres do not occur naturally.
3. Nylon is a synthetic [ ] / natural [ ] fibre.
4. Cotton [ ] / Woollen [ ] clothes keep our bodies warm in winters.
5. Raincoats protect us from rain [ ] / cold [ ].

C. Write (T) for a true statement and (F) for a false one:

1. Clothes are not necessary for us. [ ]
2. We wear clothes to cover and protect our bodies. [ ]
3. A fibre is a thread from which clothes are made. [ ]
4. Cotton and wool are natural fibres. [ ]
5. We get silk from sheep. [ ]

D. Find out the mistake and rewrite each of the given statements correctly :

1. Clothes protect our bodies from the heat of the gas.

____

____

____

2. We wear half-sleeved clothes to protect ourselves from insects.

____

____

____

3. We wear party clothes when we go to school.

____

4. A housewife wears a special uniform.

Get Hands-on Experience...

Form a club and collect old clothes. Donate them to the poor who are in need. Make sure clothes are in good condition and wearable!

Hands-on Learning... 

Collect the left-over bits of the fabrics of various types. Paste them in your album. Write the name of the fibre each fabric is made of.

12 Force, Work and Energy

We Will Learn

- Force
- Work
- Machines
- Energy

Rack Your Brain

Dear children! You must have watched a cricket match. When a bowler bowls the ball towards the batsman, the batsman applies force and hits the ball which goes in a different direction. This shows that force can change the direction of a moving object. You can observe a swing moving back and forth after somebody has pushed it. Similary, when you place a magnet near an iron nail, the magnet pulls the iron nail towards it. If you throw an object upwards, you find that it always falls back to the earth's surface. It clearly shows that force, work and energy are interlinked. Let us try to understand these interconnected terms one by one.

Force

Any kind of force is really just a push or a pull. For instance - when we push something or pull it then we are applying force to it. Force can cause an object to accelerate, slow down, remain in place, or change shape. In Nature, there are two types of force, namely gravity and friction.

Force of Gravity

Gravity is the force that attracts two bodies toward each other, the force that causes apples to fall towards the ground and the planets to orbit the Sun. The objects with greater mass exert a greater force.

Force of Friction

When a ball is rolled on the ground, it moves a distance and then stops. What is the reason behind the stopping of the ball? It stops due to friction between the ground and the ball. Friction is force that holds back the movement of a sliding object. The force acts in the opposite direction to the way an object wants to slide. If a car needs to stop at a stop sign, it slows because of the friction between the brakes and the wheels. If you run down the sidewalk and stop quickly, you can stop because of the friction between your shoes and the cement. It is friction which makes us able to walk on the floor and to hold something with our hands. If there were no friction, we would not hold an object.

Ball moving on the ground

Grip of shoes

Grip of tyre

Work

In science, work is said to be done when force is applied on a body to make it move through a distance. For example, if you push a box and move it for some distance, then work is done by you. There are other good examples of work that can be observed in everyday life - an ox pulling a plough through the field, a weightlifter lifting a barbell above his head, etc.

A man pushing a box

Machines

A machine is a device which makes the work easier and reduces our effort. Some of the simple machines that we use in our daily life are as follows :

- **Lever**
- **Inclined plane**
- **Wheel** and **axle**
- **Screw**
- **Wedge**
- **Pulley.**

Lever

A lever is one of the mechanical devices. A lever consists of a beam or stick or rod. It is used for lifting weights, cutting things and opening lids.

Seesaw Scissors Bottle Opener

Inclined Plane

An inclined plane is any slope or ramp, like a wheelchair ramp or a slide. It makes it easier to lift something heavy, like a rock. The most common example of inclined plane is the board used for loading heavy barrels or sacks onto trucks.

A man loading a container onto the truck through an inclined flat board

Brainy Point

Huge blocks of stone were pushed and pulled up on inclined planes to build the Great Pyramids of Egypt.

Wheel and Axle

Wheel and axle is a simple lifting machine consisting of a rope which unwinds from a wheel onto a cylindrical drum or shaft joined to the wheel to provide mechanical advantage. The examples of wheel and axle include car, roller, bicycles, etc.

Screw

A screw can work by turning the screw into an object or a flat surface. It is used for lowering and raising things. It is also used for holding objects together.

Axe

Knife

Wedge

A wedge is a simple machine made up of two inclined planes put together. It helps to cut things.

Pulley

A Pulley is a simple machine made up of a wheel and a rope. The rope fits into the wheel and one end of the rope is attached to the load. It is used for drawing water from the well, and in the crane for lifting heavy materials.

Pulley

Drawing water from well with the help of pulley

Energy

Energy causes things to happen around us. We need energy to do our work. It is defined as the ability to do work. Energy exists in various forms like heat energy, chemical energy, light energy, electrical energy, mechanical energy, etc.

1. Heat Energy

Heat energy is one of the most important forms of energy. The Sun is the main source of heat energy. Other sources of heat energy are fuels, like wood, gas, petroleum, etc. Electricity also produces heat energy.

Sun

Burning Wood

Gas

2. Light Energy

Light is another important form of energy. The Sun is the main and natural source of light energy on the earth. Other sources of light energy include bulb, lantern, candle, electricity, etc.

Lantern

Bulb

Candle

3. Chemical Energy

Chemical energy is obtained by chemical reaction. For example, a dry cell has chemical energy in it. This chemical energy is converted into electricity.

Petroleum

Dry cells

4. Electrical Energy

Electrical energy is the energy derived from electrical potential energy. Electrical gadgets like bulb, fan, heater, air conditioner, etc. work with the help of electrical energy.

Ceiling Fan

Heater

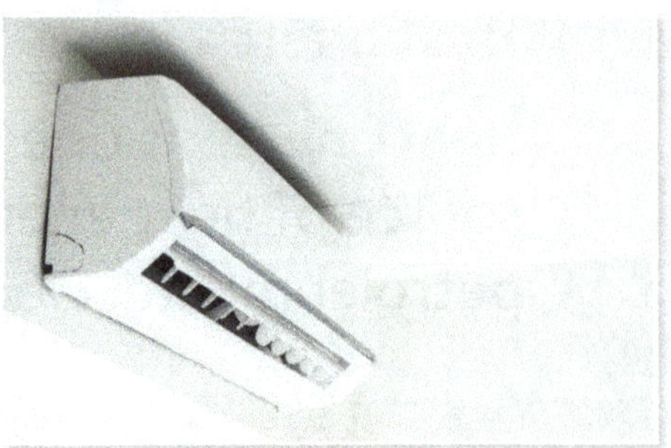

Air conditioner

5. Mechanical Energy

The energy produced by moving objects is known as **mechanical energy**. Moving air and water are used for running windmills.

Summary of the Chapter

- A push or a pull that changes the movement of an object is called force.
- The two natural forces in Nature are gravitational and frictional force.
- Energy is needed to do different kinds of work.
- Machines are used for making work easier.
- The examples of simple machines are lever, pulley, inclined plane, screw, wedge, wheel and axle.

A. Answer the following questions :

1. What do you understand by force? Explain.
2. What are the types of force?
3. What is the force of gravity?
4. What is work?
5. What is a simple machine?
6. Explain the various forms of energy.

B. Fill up the blanks :

1. ______________ and ______________ are two types of force.
2. It is ______________ that makes us able to walk on the floor.
3. A ______________ helps to cut things.
4. ______________ is the energy produced by moving objects.

C. Tick (✓) the correct answer :

1. Any kind of force ☐ /work ☐ is just a push or pull.
2. Gravity ☐ /Friction ☐ is force that tries to stop a moving object.
3. Rough ☐ /Smooth ☐ surface has less friction.
4. We need force ☐ /energy ☐ to do our work.
5. An axe is a wedge ☐ /screw ☐.

D. Write (T) for a true statement and (F) for a false one:

1. If we push or pull something then we are applying force to it. ☐
2. The earth pulls anything towards itself by frictional force. ☐
3. The seesaw is a lever. ☐
4. The Sun is the main source of heat on the earth. ☐
5. Potential energy is due to the motion of an object. ☐

E. Give one example of each of the following machines :

1. Lever

____

____

2. Inclined plane

____

____

3. Wheel and axle

____

____

4. Screw

5. Wedge

6. Pulley

Get Hands-on Experience...

Electricity is one of the important sources of energy. These days we are using electricity in excessive because of which it is getting short in supply. We should try to conserve electricity.

Design a poster to create awareness of the utility of energy. We need to use efficient lightbulbs instead of the traditional incandescent bulbs. LED (Light Emitting diode) and CFL (Compact Fluorescent Light) use 25%-80% less energy than traditional incandescent bulbs, thus saving electricity and money. They also can work 3-25 times longer than traditional bulbs.

Hands-on Learning...

Collect some pictures of simple machines (examples- lever, pulley, screw, wedge, wheel and axle) and paste them in your scrapbook.

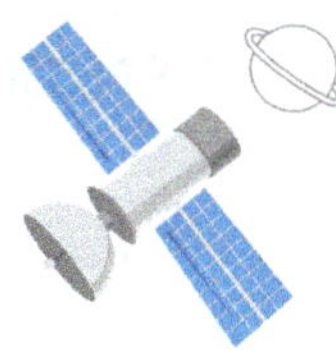

13 Measurements

We Will Learn

- Measurement of Length
- Measurement of Mass
- Measurement of Time
- Measurement of Temperature
- Measurement of Volume

Rack Your Brain

How do we know whether something is long, short, heavy or light, hot or cold, etc.? We need to measure things. The term 'measure' means to find out the certain size, amount or degree of something by comparing it with a standard unit or with an object of the known size.

Why do you compare your height with that of your friend? You want to know how much tall your friend is.

When you go to the market, you tell the shopkeeper how much rice you need.

You would like to know how long it takes to reach your school and how long it takes for a period to get over. You would also like to know how hot or cold a day is. Do you know how to measure length, mass, time and temperature?

We use different instruments to measure length, mass, time and temperature.

Let us study about the different instruments used in different measurements.

Measurement of Length

We use different measuring tools to measure length. The metre scale, the ruler and the measuring tape are commonly used tools for measuring the lengths of objects.

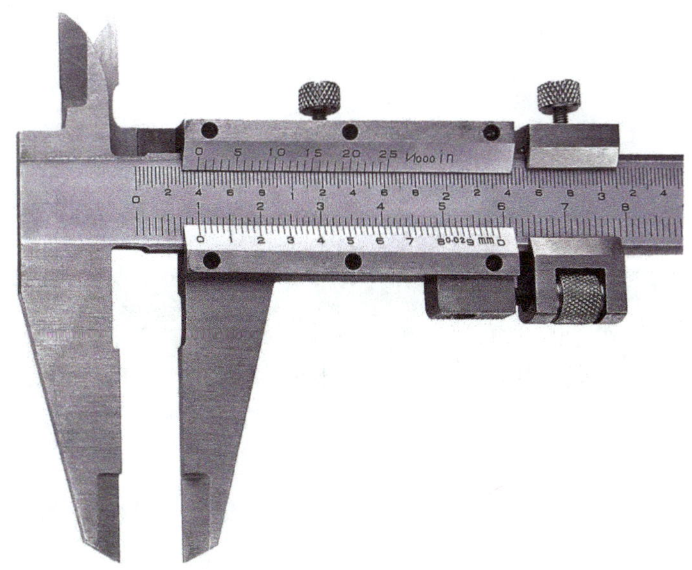

The standard unit of length is metre.

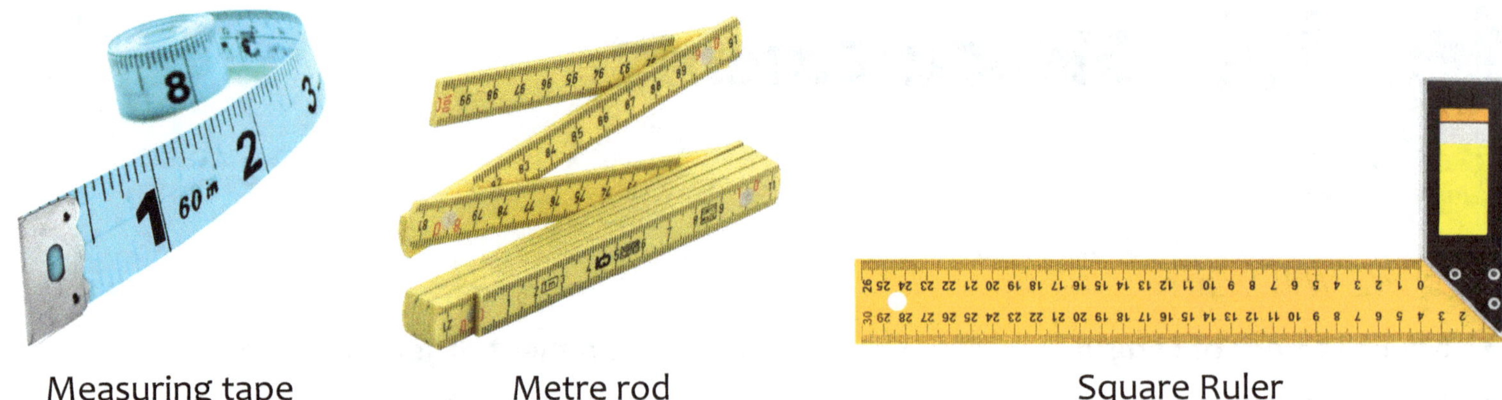

Measuring tape　　Metre rod　　Square Ruler

A cloth merchant measures his cloth with a rod. This rod is called a metre rod. How do you measure lines? We measure lines with a ruler or a scale.

You must have seen a tailor cutting a piece of cloth for stitching. He often measures cloth with a measuring tape.

The length of an object can be measured correctly using the zero mark of the scale. We often measure the length of an object by putting the zero mark of the scale on its either end.

You will see in some scales the ends may be broken or the zero mark may not be clear. So in such cases, we do not take measurement from the zero mark of the scale. We use any other full mark of the scale by subtracting this mark from the end. For example, the reading at one end of the scale is 1.0 cm and at the other end is 6.0 cm.

Therefore, the length of the object is (6.0-1.0) cm = 5.0 cm. The correct position of the eye is also very important for taking the correct measurement. We should read the measurement carefully.

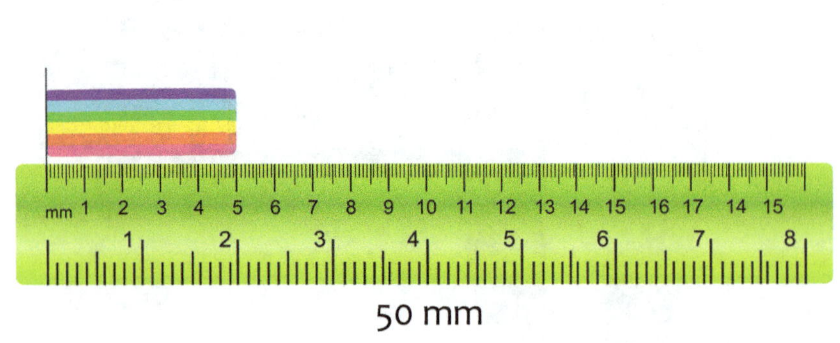

50 mm

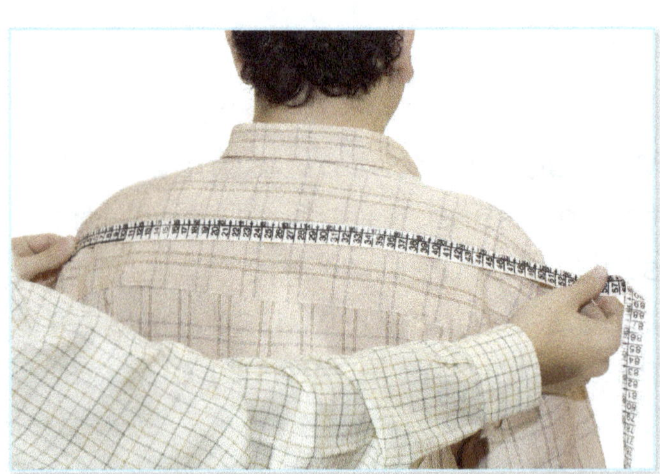

How to measure length

Measurement of Mass

Kilogram is the standard unit of mass. The mass of an object is measured with the help of the instrument like a balance or a weighing machine.

The things to be weighed are put in one pan and the weights in the other pan.

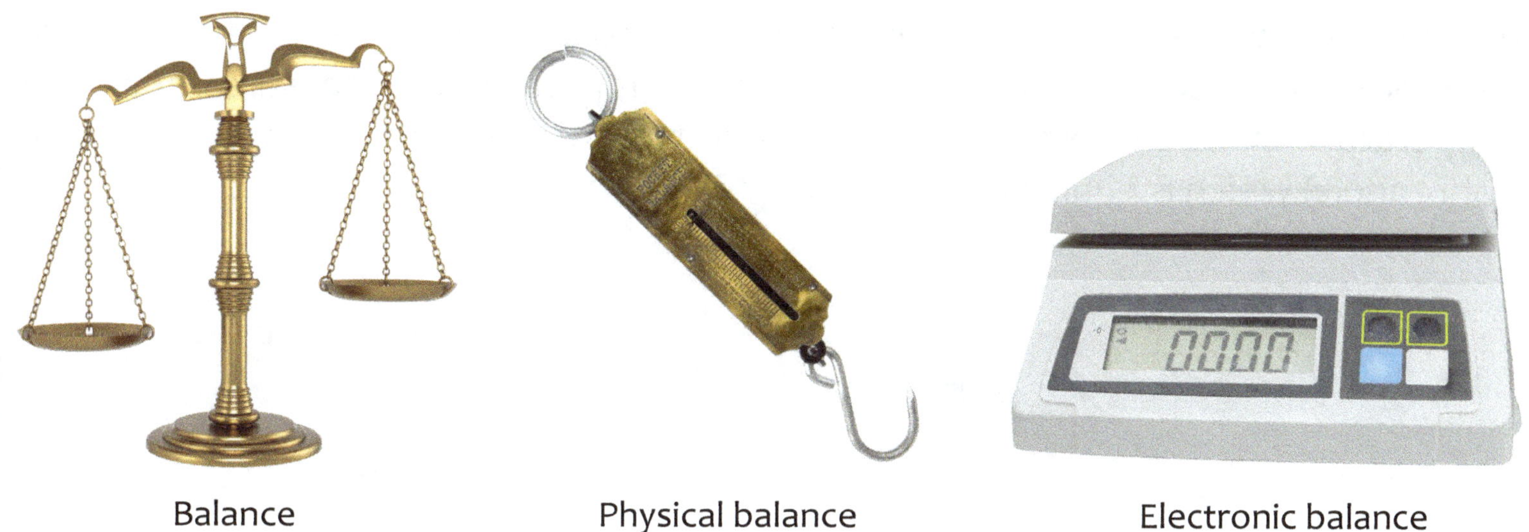

Balance Physical balance Electronic balance

Measurement of Time

We measure time with the help of a watch or a clock. It is measured in seconds, minutes and hours. The standard unit of time is second.

You must have seen a clock. There are three hands in most of clocks.

The short and thick hand reads hours.

The long and thick hand reads minutes.

The thin and long hand reads seconds.

Nowadays digital clocks are used. They work with the electric current supplied by a cell. They directly display time in numbers.

Wrist watch

Digital clock

Measurement of Temperature

The temperature of a body tells us how hot or cold it is. The temperature of a body can be measured by using an instrument called the thermometer. The standard unit of temperature is Kelvin (°K). However, we often measure temperature in degree Celsius (°C).

You must have seen a clinical thermometer. Whenever we have fever, a clinical thermometer is used for measuring our body temperature. It measures our body temperature in Fahrenheit (°F). The temperature of a healthy person is 98.6 °F (Fahrenheit) which is equivalent to 37°C.

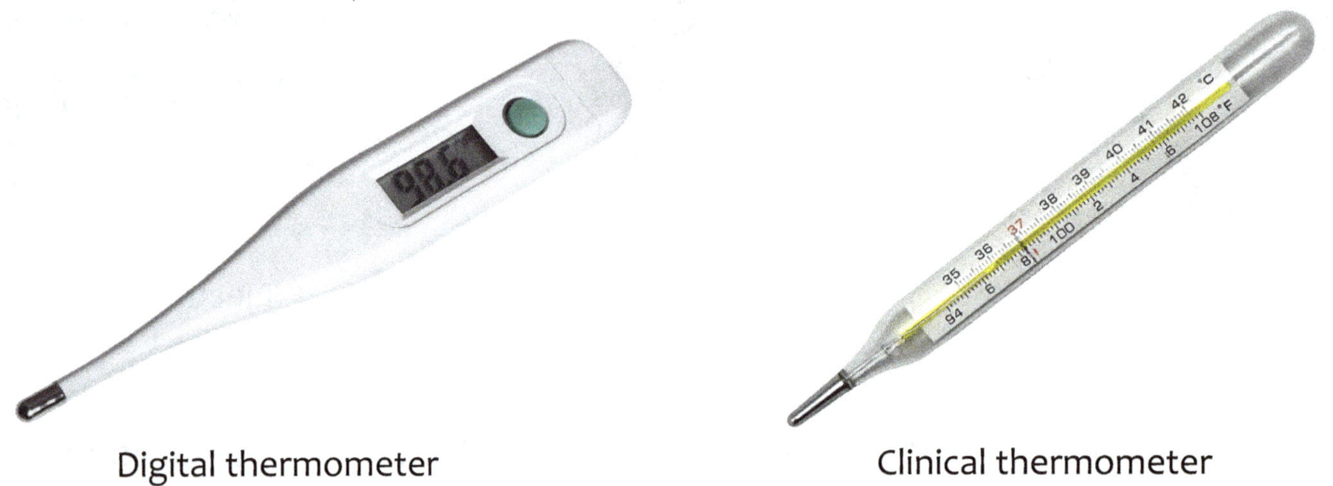

Digital thermometer

Clinical thermometer

Measurement of Volume

Volume is the total capacity of an object. It can also be considered as the total space occupied by an object.

Liquids and gases are measured by volume.

The standard unit of volume is litre.

Liquids are measured by using the vessels of different capacities.

Different types of vessels measuring volumes

Summary of the Chapter

- We use standard units to measure things.
- We use different instruments to measure length, mass, time and temperature.
- The metre scale, the ruler and the measuring tape are commonly used for measuring the lengths of objects.
- The standard unit of mass is kilogram.
- Time is measured in seconds, minutes and hours.
- The temperature of an object or a body tells how hot or cold it is.
- The temperature of a body is measured by using an instrument called the thermometer.
- The temperature of a healthy person is 98.6° F which is equivalent to 37°C.

A. Answer the following questions :

1. What are different tools to measure length?
2. What is the standard unit of length?
3. What is the standard unit of mass?
4. What is the standard unit of time?
5. What is a thermometer?
6. What is the standard unit of temperature?
7. What is meant by 'volume'?
8. What is the standard unit of volume?

B. Write (T) for a true statement and (F) for a false one :

1. The metre scale is used for measuring the length. ☐
2. The correct position of the eye is not important to take measurement. ☐
3. The mass of an object is measured by volume. ☐
4. Digital clocks display time in number. ☐

C. Fill up the blanks :

1. The standard unit of length is ____________.

2. Time is measured with the help of a ______________.
3. The temperature of a body tells us how ______________ or ______________ it is.
4. The clinical thermometer is used for measuring our body ______________.
5. Liquid and gases are measured by ______________.

D. Match the following columns :

	Column A		Column B
1.	Length	a.	litre
2.	Mass	b.	degree Celsius
3.	Time	c.	kilogram
4.	Temperature	d.	metre
5.	Volume	e.	second

Get Hands-on Experience...

Rain is measured with the help of an instrument called rain gauge. Make your own rain gauge.

Take a plastic bottle. Cut the neck of the bottle. Use the ruler to mark a scale on the side of the bottle. Set up your rain gauge in an open ground. Fix it firmly in the ground so that it may not be blown over by the wind. Check the amount of rain that falls every day.

Hands-on Learning...

Take a measuring tape measure the heights of your friends and write them in the given box:

Friend's Name	Height

14 Matter and Its States

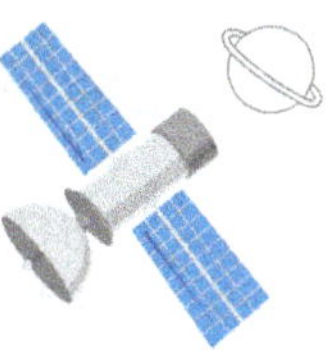

We Will Learn

- Matter and Its States
- Solute, Solvent and Solution
- Separation of Substances

Rack Your Brain

Dear children! Everything around us is made of matter. Matter is anything that has mass (weight) and takes up space. We can understand it by an example. Sanju filled a glass bowl up with water; Birju dropped a few pebbles in the water bowl, and the water spilled out of the bowl. Why did the water overflow? The pebbles raise the level of water high in the bowl and it spilled out. Let us learn more about matter in this chapter.

Matter and Its States

Matter is defined as anything that has mass and takes up space. Matter is found in three major states:-

- **Liquid :** Water, juice, milk, etc.
- **Solid :** Table, chair, book, etc.
- **Gas :** Oxygen, nitrogen, water vapour, etc.

Let us read about the states in detail.

Liquid

Solid

Gas

Liquids

Liquids are substances in which particles are not very closely packed.

- Liquids are hard to hold or grab.

- Liquids flow downward.
- Liquids can change shape depending on the container they are in. They take the shapes of the containers in which they are stored.
- The volume of a liquid never changes.

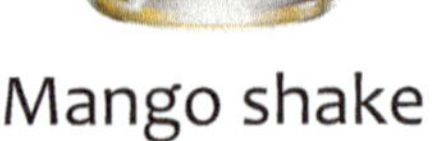
Mango shake

Apple juice

If you put liquid in a container, the molecules will slide around to fill that container

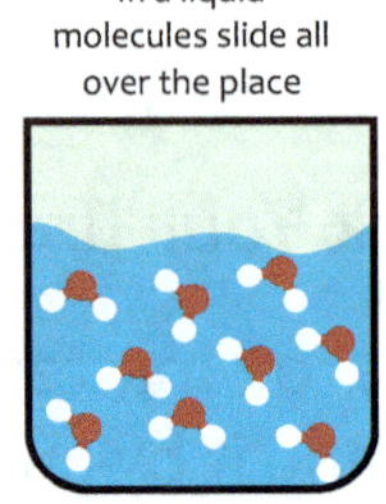
In a liquid molecules slide all over the place

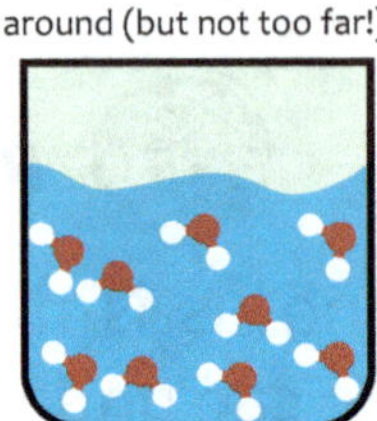
They continue to constantly slide around (but not too far!)

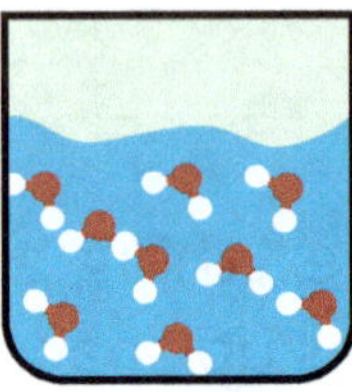

Book

Stone

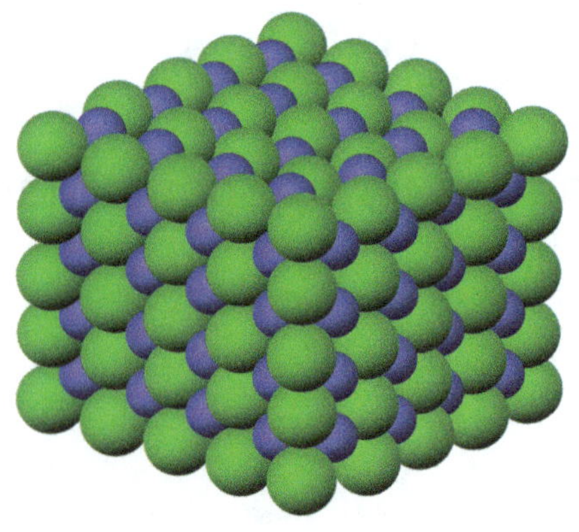
Molecules of solids

Solids

The matter that is composed of atoms packed tightly together is known as solid.

Solids are easy to control.

Anything you can grab or hold is a solid.

Solids have definite shapes and volumes.

The shape and volume of a solid will not change unless you break a bit off.

Gases

Molecules of Gases

- Gases are substances in which particles are very loosely packed.
- Gases are very hard to control.
- Gases are all around us and most of them are invisible.
- Air is a mixture of different gases.
- When put in a container, gases change their shapes and volumes to fill up the space.
- Gases will always fill the containers in which they are put in.

Change of States

Matter can be changed from one state to another, but can still be the same substance. A change of state is a physical change from one state of matter to another. For example, from solid to liquid or from liquid to gas. Let us read more about the change of states.

Heating

Heat can change solids into liquids and liquids into gases.

If ice (solid) is heated, it changes into water (liquid). This change is called melting.

Water (liquid) can change into water vapour (gas). This is called evaporation.

If water (liquid) is heated until it boils, it changes into water vapour (gas) very quickly. Water boils at 100°C.

Cooling

Cooling changes a gas into a liquid, and a liquid into a solid. If water vapour (gas) is cooled, it changes into water (liquid). This change is called condensation.

If water (liquid) is cooled, it changes into ice (solid). This change is called freezing. Water freezes at 0°C.

Gas —Cool→ Liquid —Cool→ Solid

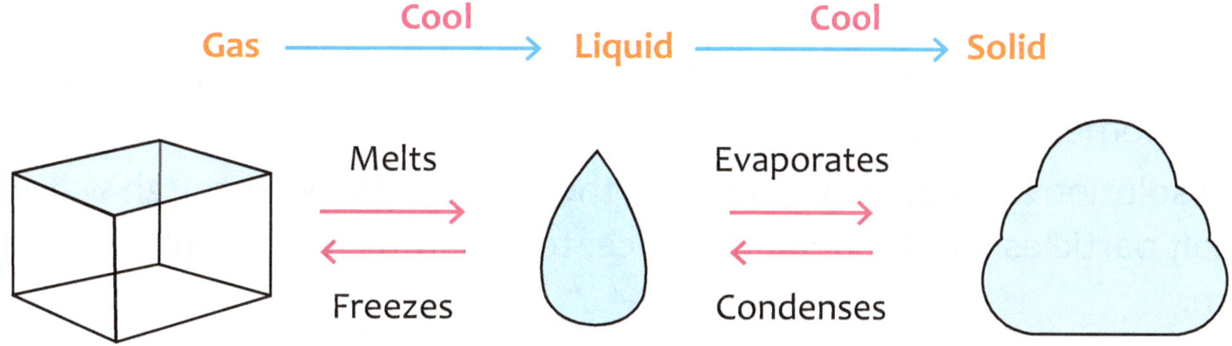

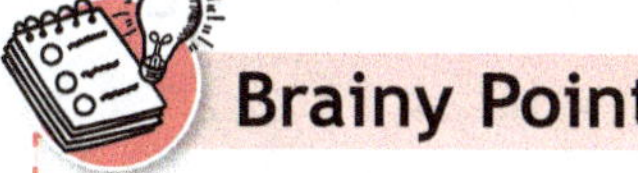

Brainy Point

Molecules are made up of smaller particles called atoms, which are combined in different ways to form molecules.

Let's now learn to change the states of solids and liquids.

Heat melts a solid and turns it into a liquid. Cooling freezes a liquid into a solid.

Solute, Solvent and Solution

Take a glass of water and put one teaspoonful of salt in it. What do you observe? You will notice that it gets dissolved in water. The solids which get dissolved in water are known as solutes. For example, salt and sugar. The liquid in which the solute gets mixed is known as the solvent. The liquid that we get by mixing solute and solvent is known as the solution.

For example: we mix sugar in hot milk. The sugar is the solute, the milk is the solvent and the mixture of both is solution.

SOLUTE + SOLVENT = SOLUTION

Separation of Substances

There are two kinds of substances:

- Soluble substances
- Insoluble substances

The substances which dissolve completely in water, like salt and sugar, are called soluble substances.

The substances which do not dissolve in water, like sand, chalk, etc., are called insoluble substances.

There are many ways by which we can separate substances from a solution. Let us learn about some of them.

Take a salt solution and heat it. On heating, the water vanishes through water vapour and the salt particles are left behind. Hence, to separate soluble substances we can heat them.

Insoluble substances cannot be separated by heating. They can be separated by three processes – filtration, sedimentation and decantation. You will learn about these methods in higher classes in detail.

Brainy Point

Aerated drinks have carbon dioxide dissolved in them. Carbon dioxide gives drinks a fizzy taste. If the drink is left open, carbon dioxide escapes and the drink tastes flat.

Summary of the Chapter

- Anything that has weight and occupies space is called matter.
- Matter is all around us.
- Matter exists in three different states– solids, liquids and gases.
- Matter can change from one state to another by heating or cooling.
- The solid that dissolves in a liquid is called the solute and the liquid that dissolves the solute is called the solvent.

A. Answer the following questions :

1. What is matter?
2. What are the three states of matter? Give examples.
3. Write one property of each state of matter.
4. What changes occur in a liquid on heating and cooling?
5. What is a solute?
6. What is a solvent?
7. What is the difference between soluble and insoluble substances?

B. Fill up the blanks :

1. If ice is heated, it changes into water. This change is called ______________.
2. If water (liquid) is heated until it boils, it changes into ______________.
3. Most solids melt into liquid when ______________.
4. ______________ freezes into a solid when it is cooled.
5. The liquid that we get by mixing ______________ is known as the solution.

C. Tick (✓) the correct answer :

1. In solid, particles are packed very closely ☐ / loosely ☐.
2. Solids ☐ / Liquids ☐ have definite shapes and volumes.
3. Air is a liquid ☐ / gas ☐ .

4. Heating ☐ / Cooling ☐ can change solids into liquids.

5. The liquid in which solute gets dissolved is known as solvent ☐ / solution ☐.

Get Hands-on Experience...

Make a list of solid, liquid and gaseous things which you see in your house. Discuss it with your friends and teachers.

Hands-on Learning...

Design a poster with water colours, depicting the amazing varieties of solids, liquids and gases available on the Earth.

Assessment Sheet - I

A. Answer the following questions :

1. What is adaptation?
2. What are milk teeth and permanent teeth?
3. How do plants and animals depend on each other?
4. Why do animals reproduce?
5. Name various methods by which we can preserve food.
6. How does a cactus survive in deserts?
7. What is evaporation? Explain with an example.
8. Which planet is also known as the Red planet? Why?
9. Write the difference between natural and synthetic fibres.
10. What is the difference between soluble and insoluble substances.

B. Fill up the blanks :

1. The flat part of a leaf is called ____________ .
2. Animals breathe out ____________ .
3. ____________ is an essential factor of all living beings.
4. The young cockroach is known as ____________ .
5. More than half of our body-weight is ____________ .
6. The white jelly-like substance in eggs is called ____________ .
7. The food produced is in the form of ____________ .
8. Sedimentation is used for removing ____________ .

C. Tick (✓) the correct answer :

1. Any kind of force ☐ / work ☐ is just a push or pull.
2. Gravity ☐ / Friction ☐ is force that tries to stop a moving object.
3. Rough ☐ / Smooth ☐ surface has less friction.
4. We need force ☐ / energy ☐ to do our work.
5. An axe is a wedge ☐ / screw ☐ .

D. Draw, label and explain the life cycle of a butterfly :

E. Tick (✓) the correct answer :

1. Which of these is a mammal?

   a. Frog ☐ b. Cat ☐ c. Parrot ☐

2. Plants take in ____________ .

   a. oxygen ☐ b. air ☐ c. carbon dioxide ☐

3. Pulses are a source of ____________ .

   a. proteins ☐ b. roughage ☐ c. fats ☐

4. The last stage of digestion happens in the ____________ .

   a. kidney ☐ b. small intestine ☐ c. large intestine ☐

5. The tiny tubes on leaves are called ____________ .

   a. cells ☐ b. stomata ☐ c. veins ☐

Assessment Sheet - 2

A. Answer the following questions :

1. What is weather ? What is the role of the Sun in the change of weather ?
2. Explain the various forms of energy.
3. What is a simple machine ?
4. What are the three states of matter ? Give examples.
5. Why do we need clothes ?
6. Which nutrients are known as body-building foods?
7. What is a dental floss? What is its use?
8. What do you mean by first aid?
9. How does a snake adapt to move without legs?
10. Why is it necessary to maintain balance in Nature?

B. Fill up the blanks :

1. Sedimentation is used for removing ______________ .
2. ______________ and ______________ are two types of force.
3. Most accidents are caused by ______________ .
4. Most solids melt into liquid when ______________ .
5. Clouds, dew and fog are the examples of ______________ .
6. Vitamins and minerals protect our bodies from ______________ .
7. ______________ are used for tearing flesh.
8. Our ______________ helps to push food into the back of our throat.
9. Fish ad crabs have ______________ to breathe in water.
10. Mammals breathe through their ______________ .

C. Write (T) for a true statement and (F) for a false one :

1. Cotton and wool are natural fibres. [ ]
2. Mercury is also called evening star. [ ]
3. First aid can save a person's life. [ ]
4. The moving air is called breeze. [ ]

5. The Sun is the main source of heat on the earth.

6. A leaf is made up of small sections called veins.

7. We get silk from sheep.

D. Explain the following processes with the help of examples :

1. Evaporation

2. Freezing

3. Chlorination

4. Sedimentation

5. Decantation

E. Describe the following with the help of examples :

1. Omnivores
2. Carnivores
3. Herbivores
4. Scavengers
5. Parasites